Studies in Computational Intelligence

Volume 1094

The series "Studies in Computational Intelligence" (SCI) publishes new developments and advances in the various areas of computational intelligence—quickly and with a high quality. The intent is to cover the theory, applications, and design methods of computational intelligence, as embedded in the fields of engineering, computer science, physics and life sciences, as well as the methodologies behind them. The series contains monographs, lecture notes and edited volumes in computational intelligence spanning the areas of neural networks, connectionist systems, genetic algorithms, evolutionary computation, artificial intelligence, cellular automata, self-organizing systems, soft computing, fuzzy systems, and hybrid intelligent systems. Of particular value to both the contributors and the readership are the short publication timeframe and the world-wide distribution, which enable both wide and rapid dissemination of research output.

Indexed by SCOPUS, DBLP, WTI Frankfurt eG, zbMATH, SCImago.

All books published in the series are submitted for consideration in Web of Science.

Loveleen Gaur · Ajith Abraham
Editors

Role of Explainable Artificial Intelligence in E-Commerce

Springer

Editors
Loveleen Gaur
School of Computer Science
Taylor's University
Subang Jaya, Malaysia

Ajith Abraham
Bennett University
Greater Noida, India

ISSN 1860-949X ISSN 1860-9503 (electronic)
Studies in Computational Intelligence
ISBN 978-3-031-55614-2 ISBN 978-3-031-55615-9 (eBook)
https://doi.org/10.1007/978-3-031-55615-9

This Springer imprint is published by the registered company Springer Nature Switzerland AG
The registered company address is: Gewerbestrasse 11, 6330 Cham, Switzerland

Paper in this product is recyclable.

Contents

Contributors

Anam Afaq Asian Business School, Noida, India

Ruqqaiaya Begum Department of Computer Science and Engineering, Vardhaman College of Engineering, Shamshabad, India

Meenu Chaudhary Noida Institute of Engineering and Technology, Greater Noida, India

Loveleen Gaur Taylor's University, Subang Jaya, Malaysia; University of South Pacific, Suva, Fiji

Richa Goel Department of Economics and IB, SCMS Noida, Symbiosis International University, Pune, India

S. Faizal Mukthar Hussain Department of CSE, Mohamed Sathak Engineering College, Kilakarai, India

M. A. Jabbar Department of CSE-AIML, Vardhaman College of Engineering, Hyderabad, India

R. Karthikeyan Department of CSE (AI&ML), Vardhaman College of Engineering, Hyderabad, India

Chitra Krishnan Symbiosis Centre for Management Studies, Symbiosis International Deemed University, Noida, India

Jasmine Mariappan University of Technology and Applied Sciences, Ibra, Oman

Jagini Naga Padmaja Department of Computer Science and Engineering, Vardhaman College of Engineering, Shamshabad, India

Garima Sahu Amity International Business School, Amity University, Noida, India

Aatam Parkash Sharma Department of Marketing and Supply Chain Management, School of Business Studies, Central University of Jammu, Jammu and Kashmir, India

Naresh Kumar Sharma Department of Marketing and Supply Chain Management, School of Business Studies, Central University of Jammu, Jammu and Kashmir, India

Neeru Sidana Amity School of Economics, Amity University, Noida, India

Gurinder Singh Amity International Business School, Amity University, Noida, India

Koti Tejasvi Department of Computer Science and Engineering, Vardhaman College of Engineering, Shamshabad, India

Kavita Thapliyal Amity International Business School (AIBS), Amity University, Noida, India

Manjul Thapliyal Visions Ahead Foundation, Noida, India

V. Lokeshwari Vinya Department of Computer Science and Engineering, Vardhaman College of Engineering, Shamshabad, India

Introduction to Explainable AI (XAI) in E-Commerce

Meenu Chaudhary, Loveleen Gaur, Gurinder Singh, and Anam Afaq

Abstract In the ever-evolving landscape of technology, companies strive for innovation to maintain competitiveness. Artificial Intelligence (AI) has permeated every sector, and in the realm of e-commerce (EC), its impact is notably evident. Various AI applications, such as recommendation systems, fake filters, and fraud detection, have greatly benefited the EC industry. However, a lingering issue is the challenge of understanding and explaining the outcomes generated by AI algorithms, which, in turn, affects their trustworthiness. In addressing this concern, there's an ongoing discourse regarding the ethics and privacy implications of AI, prompting additional research endeavors. The objective is to enhance the trustworthiness and ethical standing of AI systems. This has led to the resurgence of Explainable AI (XAI), a domain focused on making AI results more comprehensible to users. The prevailing challenge lies in the fact that existing technologies often fall short in providing detailed explanations of how algorithms arrive at specific results or recommendations. Specifically in e-commerce, where decisions often demand immediate action, the integration of XAI systems becomes crucial. These systems aim to provide instant justifications, filling the gap left by current technologies that struggle to offer thorough explanations of the decision-making process behind AI-generated results or recommendations.

Keywords E-commerce · Explainable AI · XAI · Artificial intelligence

---

M. Chaudhary (✉)
Noida Institute of Engineering and Technology, Greater Noida, India
e-mail: chaudharymeenu1991@gmail.com

L. Gaur
Taylor's University, Subang Jaya, Malaysia

University of South Pacific, Suva, Fiji

G. Singh · A. Afaq
Asian Business School, Noida, India

L. Gaur and A. Abraham (eds.), *Role of Explainable Artificial Intelligence in E-Commerce*, Studies in Computational Intelligence 1094,
https://doi.org/10.1007/978-3-031-55615-9_1

1 Introduction

EC refers to the online buying and selling of products or services, driven by the growing demand for online services and the competitive advantage it provides to businesses. However, this e-business practice presents challenges due to its integration with readily available data and the rapidly advancing, cost-effective information technology [1]. Businesses must continually adapt their strategies to meet evolving customer needs, with the latest innovation being artificial intelligence (AI) [2]. AI's capacity to comprehend external data, learn from it, and apply those learnings to achieve specific goals is transforming the landscape of EC. In the context of EC, AI involves the application of AI methods, systems, instruments, or algorithms to facilitate online buying and selling activities. AI aids businesses in decision-making by analyzing vast volumes of data quickly and with minimal human error [3, 4]. The advantages of AI for businesses include rapid pattern identification, visualization of analysis, market and competitor insights, leading to increased revenue, cost restructuring, and improved efficiencies. It allows businesses to gain a competitive edge by offering customized services tailored to the specific needs of customers. AI has become ingrained in users' lives, influencing daily decisions, from movie and product recommendations on platforms like Netflix and Amazon to personalized advertisements in Google search results and friend suggestions on Facebook. The term 'artificial intelligence' was coined by McCarthy, a Stanford University professor, at the Dartmouth Conference, envisioning systems that would develop intelligence akin to humans in the future [5]. Although early hypotheses, such as those by Firschein and Coles in 1973, faced limitations due to technological constraints, subsequent developments have seen the realization of products like automated industrial automata and linguistic interpreters [6, 7].

AI technology encompasses intelligent devices or software capable of human-like cognition [8, 9]. Its applications extend across various disciplines, including education, marketing, consumer behavior analysis, past purchase record evaluation, and competitor analysis [10]. There are two types of AI: weak AI, programmed for complex tasks but requiring human intervention, and strong AI, possessing consciousness like humans and capable of autonomous decision-making [11, 12]. Despite the effectiveness of AI algorithms in delivering results and predictions, they often lack transparency, especially in Machine Learning (ML) algorithms [13]. This opacity poses challenges in understanding their internal dynamics, highlighting the risks associated with relying on a system that cannot explain itself. To address this challenge, Explainable Artificial Intelligence (XAI) emerges as a solution, aiming to develop techniques that generate more transparent AI models while ensuring excellence in performance.

2 XAI

XAI is a field of study that pursues to make the results of AI more comprehensible to people. Van Lent [14] coined the term in 2004 to describe the system's capacity to describe the conduct of artificial intelligence-controlled entities in simulation applications for games. Though the term is comparatively new, the matter of explainability dates to the mid-1970s, when scholars started to investigate expert system explanations. Nonetheless, AI has reached a tipping point due to the phenomenal advancement of ML, and the rate of advancement toward resolving such issues has slowed down. AI research has transferred its emphasis to the implementation of predictive models and algorithms, whereas the capacity to explain decision-making processes has fallen behind.

Academics and practitioners have recently refocused their attention on XAI. Using Google Trends, Fig. 1 depicts the notable revival of XAI research interest. The recent revival of this research topic is entirely due to the persistent permeation of AI-ML across sectors and its significant effect on complex decision-making, unable to provide specific details regarding the reasoning behind specific decisions, recommendations, or measures initiated by it. The cultural, moral, and legislative pressures mandate the creation of new AI methods that can explain and clarify decisions.

Technically, there is no widely acknowledged definition of explainable AI. Rather than a formal technical concept, the term XAI typically denotes the action, initiatives, and endeavors undertaken in response to concerns about AI transparency and trust. According to DARPA [15], the objective of XAI is to "produce more explainable models while maintaining a high level of learning performance (prediction accuracy); and enable human users to understand, trust, and effectively manage the emerging generation of artificially intelligent partners." According to FAT [16], enabling explainability in ML seeks ensuring that algorithmic decisions and the data underneath can be conveyed to decision-makers and other related stakeholders using non-technical language.

The terminology "explainable AI" characterizes an AI model, its anticipated influence, and any possible prejudices. It assists in identifying model precision, impartiality, transparency, and outcomes in AI-powered decision-making. XAI is essential for the business to establish trust and confidence before deploying AI models. It also facilitates the adoption of an accountable approach to AI advances within an organization.

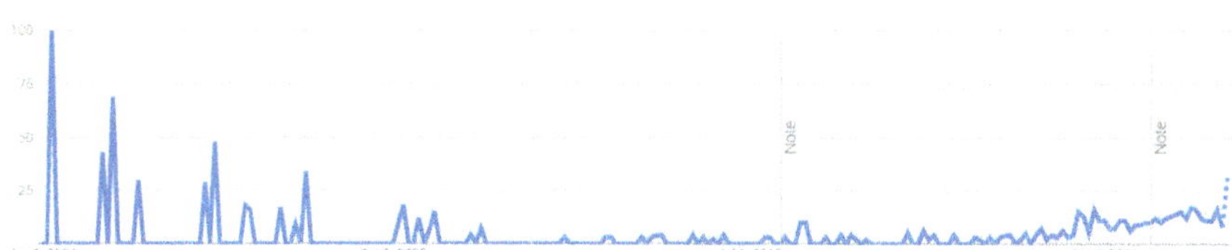

Fig. 1 Google trend of "explainable artificial intelligence" term

With more complications in AI, it becomes increasingly difficult for humans to understand and retrace the algorithm's reasoning. The entire calculation process is transformed into what is commonly known as an unintelligible "black box." These models are referred as blackbox as the engineers or data scientist who constructs the AI algorithm are unable to explain the reason for particular result. The problem arises when an algorithm produce an unbiased result, it can be used to make further predictions which can worsen the performance of the algorithm. Henceforth, reliability becomes an issue.

Learning how an AI-enabled system generated a particular output offers numerous benefits. Explainability ensures the technical as well as non-technical decision-makers that AI is giving the result as intended. It may be required to satisfy regulatory standards, and it may be crucial for allowing those affected by a decision to challenge or change that outcome.

During our literature review, we did a linguistic search to find and classify terminologies that are related to XAI across research communities. The goal is to comprehend how researchers address explainability and to identify the important concepts that make the definition of this concept. The word cloud formed provides an instant understanding on XAI scope and facilitation on the related concepts (Fig. 2).

In Table 1, the commonly related concepts of XAI can be used to contextually define the studied field.

Fig. 2 Word cloud

Table 1 Related terminologies of XAI

Interactive artificial intelligence	Interactive AI is a subset of XAI that enables the user to interact with the device in use
Transparent AI	A form of transparent artificial intelligence is essential for establishing trust between humans and the algorithms that work to assist humans in completing a variety of tasks and obtaining necessary solutions

3 Gaining an Edge Through AI

Throughout the value chain, AI technologies can provide a variety of benefits, spanning from planning to administrative tasks. Widespread data-generation tools have made it imperative for companies to device new strategies and practices that offer managers with timely and accurate information for decision-making purposes. AI can automate processes, gaining analytics from historical data, and giving customer or market statistics through program-based algorithms to aid in managerial decision-making [17]. AI can also provide benefits unrelated to labour replacement. Among these are predictive maintenance tools, optimising documentation, analysing the impact of weather on the supply chain, etc. [18]. Nonetheless, businesses must bear in mind that, usually, transitioning to a fully automated process may necessitate significant process-wide changes, as opposed to merely modifying specific process components.

Seventy percent of those surveyed deemed AI to be the most advantageous instrument for personalization. Moreover, 52% of e-commerce (EC) executives believed that AI aids in predicting the demand for a product [19]. The establishment of pricing and advertising strategies, the enhancement of service to customers and client satisfaction, AI-based routing software for faster deliveries, are further fascinating applications of artificial intelligence.

AI-enabled systems can sift through vast quantities of data and identify consumer behaviour patterns [20]. This provides the users with more precise data on the market"s requirements and demands. AI is continuously modifiable and adaptable based on business requirements. For instance, a well targeted marketing campaigns can be designed with AI by giving precise suggestions to consumers based on the consumer's search, purchase, and shopping cart behavior.

Nonetheless, as depicted in Fig. 3, the utmost prevalent use of AI is in customer service, which accounts for 48% of all applications. This is followed strongly by the business's quality control and management of inventory implementations, each of which accounts for 47% [21]. This form of the technology enables businesses around the globe to continue to personalize the availability of content to customers and adjust their prices based on their behavior. Finally, businesses can ensure online security by employing AI as a system for detecting deception.

4 Artificial Intelligence in eCommerce: Benefits

1. Customer-oriented visual search

Dissatisfaction among customers for EC experiences due to irrelevant displayed product result decreases customer loyalty. AI uses natural language processing to refine, contextualize, and enhance search results for online consumers in order to combat this issue. Additionally, it enables visual search and product-matching capabilities. Additionally, AI facilitates the discovery of complementary products and

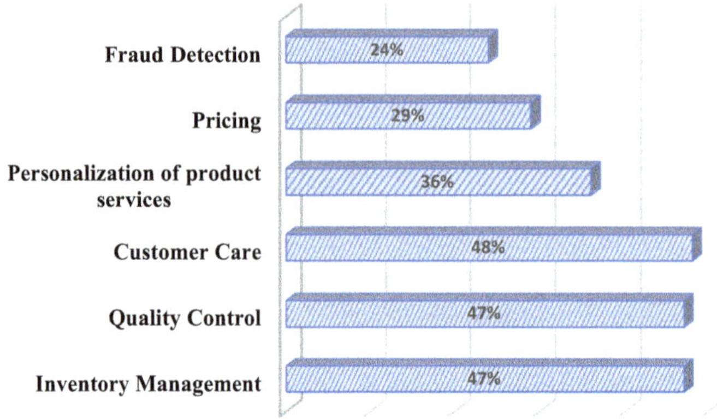

Fig. 3 AI use in E-commerce. *Source* [21]

improves the consumer experience. Consumers can take a picture of a friend's new clothes or accessories, upload it, and then AI will make it simple for them to locate identical products in online stores. Sounds amazing, right? Amazon, for example, has a feature that enables the customers to locate the product you like, and will be identified on Amazon and provide the results that the user will likely enjoy as they are precisely what you were looking for [22].

2. Re-target prospective clients

As per Conversica, sales teams do follow-ups on approx. 67% of marketing leads. It suggests that prospective consumers who are pre-qualified and have an interest in a product are overlooked. Additionally, numerous teams are overburdened with unruly consumer data that is not of much use. Then AI becomes an absolute necessity. AI could help improve the sales cycle by customizing problem-solving solutions and crafting an effective sales message that reaches consumers at the best possible time and on the ideal platform. Many AI systems, like Siri, Alexa, etc., presently support NLP and audio input. Thus, a CRM system is able to instantly react to customer inquiries, resolve their issues, and even identify new sales opportunities.

To better comprehend, let's examine an excellent illustration—The North Face is a major online retailer that uses IBM's Watson AI solution for improved comprehension of their customers. Customers can respond verbally or in writing to queries such as "Where and when will you be using your running clothes?" in order to assist them in finding the ideal apparel. The IBM software then examines thousands of items to discover seamless matches based on real-time customer input.

3. New degree of personalisation

Boomtrain, for instance, is one of the many AI solutions that are currently available. This company analyses various means of contact to aid businesses in understanding how consumers interact online using mobile apps, websites, emails. An AI engine monitors all devices and channels to generate a unified opinion of the customer. Therefore, not just one channel, but every channel. Moreover, it helps EC retailers provide a uniform experience for customers in any platform. It will aid in sending timely, pertinent messages.

4. Chatbots and virtual assistants

User experience has emerged as the primary focus of e-commerce. In the modern age of conversational commerce, the use of AI via the application of "chatbots" is merely one way to guide the conversation. Moreover, chatbots are capable of automating order processes. In addition, they are a cost-effective method to provide 24*7 customer service, collect valuable data, and monitor behaviour. Using chatbot, e-commerce websites can increase conversion rates by personalising the user's online experience. Again, one of the most recognisable examples is Alexa, Amazon's virtual assistant. It has been incorporated into Amazon's products and those of other manufacturers. It is anticipated that virtual assistants will influence consumers' purchasing decisions and provide an innovative opportunity for e-commerce retailers.

Chatbots can have a significant impact on the customer service of e-business because they can operate 24/7, assist consumers in locating products, have accessibility to the most current supply chain information, comparison of multiple products, and instruct customers regarding the way to pay. This computer program reduces labor costs because employees will no longer devote time to answering recurrent queries. Rather, the chatbot is capable of handling both basic and complex interactions, it helps the customer reach the appropriate personnel [23].

Though, chatbots does not comprehend human sentiments during interactions and may give incorrect responses, resulting in dissatisfied consumers. Additionally, chatbots start the procedure by requesting personal information from the customer, which can generate the perception of a cyber threat among consumers, particularly in case the performance is inadequate.

Currently, a vast array of practical applications demonstrates the viability of this system. Replika, for example, an AI-powered personal chatbot assistant; their ML-based chatbot emulates the conversational style of users and thus is able to offer them emotional assistance.

5. Improve recommendations for customers

Since the mid-1990s, establishments like Netflix, Spotify, YouTube, Amazon Prime have utilised recommendation engines. However, additional types of EC can also take advantage of this mechanism. The necessity of filtering huge amounts of data and provide users with tailored suggestions (of their interest) when seeking a product or service within an extensive data set led to the development of these systems.

With AI, brands can predict consumer behaviour and demand more logically and competently, as well as providing relevant and useful recommendations. Starbucks is a prime example, as it employs AI to analyse all the data it has collected in order to provide more personalised recommendations. The algorithm deliberates the customer's data, preferences, purchase history, third-party data, and contextual information. 30% of Amazon's page views are directly attributable to recommendation engines, according to studies [24]. Moreover, YouTube reported in 2010 that sixty percent of homepage views were for recommended videos. Constantly, machine learning algorithms are incorporated into recommendation engines.

ML consists of automatic analysis of data, pattern recognition, and decision-making, with minimum human intervention. As stated by Sorbán [25], ML algorithms employed by recommendation engines are categorised based on their filtration technique.

- Content-based filtering: This method assigns descriptors to each database item and constructs user profiles based on their input. It doesn't need a huge dataset; however, its advantages are less, given that only similar items are recommended.
- Collaborative-based filtering: It generates suggestions for consumers by analyzing commonalities and connections between consumers and figuring out the way in which user interact with system-managed items. This process was employed by enterprises for standard recommendation.
- Knowledge-based filtering: Current recommendation systems are not confined to sole application of the filtering techniques. In place of this, usage of diverse techniques and filtering systems is functional; these recommendation engines are commonly referred to as hybrid recommendation engines.
- Filter fake reviews

Sadly, fake evaluations have become a problem for online retailers and EC sites. As per Dimensional Research, 90% respondents indicated that positive online evaluations impact their purchasing decisions. Additionally, 86% stated that negative online reviews influenced their purchasing decisions. However, as we all know, there are competitors, bots, and others who publish false reviews. So, how can AI solve this issue? Today, many EC companies employ AI to combat astroturfing by placing a greater emphasis on authentic and useful reviews. Amazon, for instance, employs AI against fraudulent product evaluations and star rating inflation. AI emphases on the prominence and weight of authentic consumer testimonials and promotes them. AI additionally takes into account reviews that other users have marked as helpful.

7. Automation

AI does not imply that robots will take over the world. Numerous individuals fear that in the near future, all work will be performed by machines. No, actually! Using technology and algorithms, robots allow retailers to provide precisely what customers want, when they want it. As the e-commerce industry continues to develop, and with it, the number of customers, automation has become a necessity for e-commerce stores and among their foremost investment priorities. In addition, it is common knowledge that as an eCommerce business develops, the number of repetitive tasks also increases. Anything from launching new products across different platforms to sales forecasting and offering discounts to loyal consumers can be handled by robots [26].

As reported by Martechtoday, Ocado intended to expend $232 million this year on its robotics and automation system, Amazon allocated $5 billion to expenditures on technology in India alone, and Alibaba is contemplating financing $15 billion in R&D over the course of three years.

8. Fraud detection and prevention

E-commerce companies are rapidly facing issues regarding fraudulent activities. Dealing with these challenges is leading to losses both financially and reputationally [27]. Conventional fraud detection techniques frequently use rule-based systems and manual investigations, which can be time-consuming, expensive, and error-prone [28]. Using AI-based solutions for fraud detection and prevention has drawn more attention in recent years. Moreover, the black-box nature of many AI algorithms can make it challenging to understand how judgements are made, which is one of the difficulties with employing AI in fraud detection [29]. Explainable AI (XAI), which offers insights into how AI systems make decisions, can assist in resolving this problem. The causes of fraud can be found with the aid of XAI, which can also clearly explain how these causes are used while making decisions. Prioritising the most critical elements for fraud identification and prevention is one of the key advantages of employing XAI in fraud detection. E-commerce companies can concentrate their efforts and resources on the areas that are most likely to have the biggest impact by determining which factors are most predictive of fraud [30]. The success of other actions, such as modifying fraud rules or adopting new fraud prevention technologies, can also be assessed using XAI.

The capability of XAI to recognise specific user actions that are suggestive of fraud is another advantage. In order to find trends and abnormalities that can point to fraud, XAI algorithms can evaluate vast volumes of data from numerous sources, including transaction data, user activity data, and social media data [31]. E-commerce companies can improve their overall fraud detection capabilities and create more efficient fraud protection methods by comprehending these tendencies [32]. Applying XAI to fraud detection and prevention can aid e-commerce companies in reducing losses brought on by fraud, fostering customer trust, and enhancing customer experience. XAI can also help boost accountability and lower the possibility of bias or errors in AI-based fraud detection systems by offering concise and transparent explanations of how choices are made.

5 Need of XAI

The extensive customization of advertisements using AI technologies can impact consumer preferences and purchasing behaviors [12]. While AI provides decision-makers with crucial insights to assess strengths and weaknesses, the perceived value and application scope may lead retailers into having unrealistic expectations. Despite the continuous influx of investments and the development of more potent AI systems, the implementation of troubleshooting and oversight procedures lags the ambitious plans.

Moreover, the data used by AI systems to analyze consumer behavior tends to be unidirectional, primarily sourced from customer interactions and feedback. According to Khrais and Azizi [12], this data is often narrow and non-dynamic, limiting its ability to offer a comprehensive understanding of consumer behavior.

Understanding and rationalizing predictions made by AI systems have been a challenge since the 1970s. While recent advancements focus on designing more precise algorithms and models, the explanation of these systems' forecasts has become a significant concern. The complexity and autonomy of AI systems can lead them to make decisions with limited human intervention, giving rise to the persistent integration of artificial intelligence, especially machine learning (ML). Consequently, Explainable AI (XAI) has emerged as a crucial domain, aiming to make AI results more understandable to people. This becomes essential where current technologies fall short in providing detailed explanations of how algorithms generate specific results or recommendations.

Adabi and Berrada [33] gave four reasons for the importance of XAI.

- T*o Explain*: AI is not flawless and have given biased results in the past. By implementing XAI, end-users can rationalise unpredictability in machine outcomes or ensure the way to validate that the algorithmic result was unbiased and objective.
- *To Command*: Improved control system aids in the anticipation of unjust results. With an understanding of the operations of the system, the user can recognize susceptibilities or defects becomes easier, and the decision-maker can recognize and correct errors with greater speed.
- *To Improve*: Like the preceding point, users with a deeper understanding of the system are able to identify areas of deficiency. Thus, XAI encourages continual advancements in AI.
- *To Discover*: The generation of new knowledge is only possible when the user can understand the current AI operations. These systems may be intelligent enough to advance our understanding in the future of other disciplines, such as physics, biology, and so on.

With XAI, AI technologies will be employed in more advantageous manners, as it is more user-friendly and give more autonomy to the users over the machine for control. In addition, it addresses contemporary moral issues, as existing structures are fraught with ethical dilemmas. XAI aims to address the issues of wrong decision-making by machines where the input is from all types of data. Consequently, more studies in the field of XAI can result in making more comprehensible, dependable, and manageable systems.

AI algorithms are considered as BlackBox, which denotes that the internal operations of an e-device are either obscured or unclear decision-makers, resulting in sometimes incomprehensible but excellent solutions [34–39]. However, this also implies that if the decision-maker is unable to give justification for the prediction given by the system, difficulties will always be there whenever an unsuitable prediction is made. Therefore, there is a greater possibility that businesses will be more prone to influenced data, erroneous decisions based on results, and incorrect modelling practices in the algorithm.

6 Discussion

The demand for XAI has surged with the widespread adoption of machine learning techniques across various applications, including E-commerce [40]. XAI refers to an AI system's ability to elucidate its decision-making process in a comprehensible manner for humans [41]. In the business domain, XAI plays a pivotal role in enabling organizations to comprehend customer behavior, make informed decisions, and provide personalized experiences [42]. An essential characteristic of XAI in business is its capability to interpret and understand the data utilized by AI systems. This empowers businesses to explain the rationale behind decisions, uncover biases and inaccuracies in data and models, and avoid erroneous or biased judgments that could negatively impact clients and the bottom line [43]. Transparency and accountability are integral aspects of XAI in business, as it facilitates adherence to ethical standards and legal requirements by providing justifications for decision-making [44]. This, in turn, builds customer and stakeholder trust while mitigating legal risks.

Furthermore, XAI contributes significantly to enhancing the client experience in business. By offering individualized advice and justifications for specific recommendations, organizations can strengthen customer relationships, boost satisfaction, and foster loyalty [45]. XAI's ability to identify consumer preferences and behaviors aids in the creation of new products and services and the refinement of marketing strategies [46]. In the realm of fraud detection, XAI proves invaluable for companies seeking to prevent financial losses and protect their clients. By detecting fraudulent transactions and activities, XAI contributes to maintaining the integrity of review systems on e-commerce platforms. Algorithms under the XAI umbrella also play a crucial role in automation within e-commerce, handling tasks such as customer care, inventory management, and order fulfilment [47]. This not only enhances productivity but also reduces costs.

Moreover, XAI encompasses various components like personalized consumer suggestions, chatbots, virtual agents, and personalization [48]. Algorithms analyze customer data to provide recommendations for goods and services tailored to specific customers, fostering loyalty and contentment. A Microsoft study, for instance, found that personalized product recommendations based on XAI algorithms increased conversion rates by 30% [49]. In addition, the growing prevalence of virtual assistants and chatbots in e-commerce further highlights the impact of XAI [17]. XAI-powered chatbots and virtual assistants offer individualized customer service, responding to queries and providing recommendations [23]. This results in faster and more accurate customer care, elevating customer happiness and loyalty. XAI, by customizing each customer's purchasing experience, can suggest products, promotions, and offers based on individual preferences. While XAI offers numerous benefits to e-commerce businesses by fostering transparency and trust, challenges such as data privacy concerns and biases in algorithms must be carefully addressed [50, 51].

7 Conclusion

Explainable Artificial Intelligence (XAI) plays a crucial role in e-commerce, empowering companies to enhance decision-making processes, increase productivity, and deliver personalized customer experiences [52]. XAI systems, capable of processing vast amounts of data, offer insightful information on consumer behavior and preferences, aiding businesses in making informed decisions about product development, marketing plans, and customer support. However, it is imperative for businesses to diligently address potential biases in XAI algorithms. When trained on biased data or non-representative datasets, algorithms may develop biases. Regular auditing of XAI algorithms and datasets is essential to identify and eliminate biases, ensuring fairness. Moreover, organizations must prioritize secure and ethical management of customer data to maintain trust. Adherence to laws and guidelines, such as the California Consumer Privacy Act, is crucial in the era of growing concerns about data privacy and security [34–39, 53].

In summary, XAI serves as a powerful tool for e-commerce, enhancing decision-making processes and providing personalized experiences. However, businesses must proactively mitigate potential biases in XAI algorithms and uphold ethical and secure customer data practices. This approach allows businesses to retain customer confidence, loyalty, and fully harness the potential of XAI. Applying XAI to fraud detection in e-commerce can significantly reduce losses from fraud, foster customer trust, and enhance the overall customer experience. Additionally, XAI contributes to accountability and reduces the risk of bias or errors in AI-based fraud detection systems by offering transparent explanations of decision-making processes.

References

1. Gielens, K., & Steenkamp, J. B. E. (2019). Branding in the era of digital (dis) intermediation. *International Journal of Research in Marketing, 36*(3), 367–384.
2. Tan, F. T. C., Pan, S. L., & Zuo, M. (2019). Realising platform operational agility through information technology–enabled capabilities: A resource-interdependence perspective. *Information Systems Journal, 29*(3), 582–608.
3. Khrais, L. T., & Azizi, T. A. (2020). Analyzing consumer attitude toward mobile payment technology and its role in booming the e-commerce business. *Talent Development and Excellence, 12*.
4. Areiqat, A. Y., Hamdan, A., Alheet, A. F., & Alareeni, B. (2021). Impact of artificial intelligence on E-commerce development. In *The Importance of New Technologies and Entrepreneurship in Business Development: In The Context of Economic Diversity in Developing Countries: The Impact of New Technologies and Entrepreneurship on Business Development* (pp. 571–578). Springer International Publishing.
5. Gururaj, P. (2021). Artificial intelligence-application in the field of e-commerce. *Int. J. Res.-Granthaalayah, 9*, 170–177.
6. Firschein, O., Fischler, M. A., Coles, L. S., & Tenenbaum, J. M. (1973, August). Forecasting and assessing the impact of artificial intelligence on society. In *IJCAI* (Vol. 5, No. 1, pp. 105–120).

7. Soni, N., Sharma, E. K., Singh, N., & Kapoor, A. (2020). Artificial intelligence in business: From research and innovation to market deployment. *Procedia Computer Science, 167*, 2200–2210.
8. Shyna, K., & Vishal, M. (2017). A study on artificial intelligence E-commerce. *International Journal of Advances in Engineering and Scientific Research, 4*(4), 62–68.
9. Chaudhary, M., Gaur, L., & Chakrabarti, A. (2022, April). Comparative analysis of entropy weight method and c5 classifier for predicting employee churn. In *2022 3rd International Conference on Intelligent Engineering and Management (ICIEM)* (pp. 232–236). IEEE.
10. Rana, J., Gaur, L., Singh, G., Awan, U., & Rasheed, M. I. (2021). Reinforcing customer journey through artificial intelligence: A review and research agenda. *International Journal of Emerging Markets, 17*(7), 1738–1758.
11. Chaudhary, M., Gaur, L., Chakrabarti, A., & Jhanjhi, N. Z. (2023). Unravelling the Barriers of human resource analytics: Multi-criteria decision-making approach. *Journal of Survey in Fisheries Sciences*, 306–321.
12. Khrais, L. T. (2020). Role of artificial intelligence in shaping consumer demand in E-commerce. *Future Internet, 12*(12), 226.
13. Afaq, A., Gaur, L., Singh, G., & Dhir, A. (2021). COVID-19: Transforming air passengers' behaviour and reshaping their expectations towards the airline industry. *Tourism Recreation Research*, 1–9.
14. Van Lent, M., Fisher, W., & Mancuso, M. (2004, July). An explainable artificial intelligence system for small-unit tactical behavior. In *Proceedings of the National Conference on Artificial Intelligence* (pp. 900–907). AAAI Press; MIT Press; 1999.
15. Gunning, D., Stefik, M., Choi, J., Miller, T., Stumpf, S., & Yang, G. Z. (2019). XAI—Explainable artificial intelligence. *Science Robotics, 4*(37), eaay7120.
16. Barocas, S., Friedler, S., Hardt, M., Kroll, J., Venka-Tasubramanian, S., & Wallach, H. (2018). The FAT-ML workshop series on fairness, accountability, and transparency in machine learning. *cit. on*, 7.
17. Afaq, A., Gaur, L., & Singh, G. (2023). Social CRM: Linking the dots of customer service and customer loyalty during COVID-19 in the hotel industry. *International Journal of Contemporary Hospitality Management, 35*(3), 992–1009.
18. Wilburn, K. M., & Wilburn, H. R. (2018). The impact of technology on business and society. *Global Journal of Business Research, 12*(1), 23–39.
19. Statista. Future Areas in Which AI Will Help E-Commerce Businesses According to Decision Makers in North America and Europe in 2021. (2021). Available online: https://www.statista.com/statistics/1175063/ai-help-business-ecommerce-companiesworldwide/ (Accessed on 10 November 2021).
20. Chaudhary, M., Gangele, A., Naved, M., Gaur, L., & Singh, G. (2022, November). The function of driver categorisation in the ride-hailing industry: A study on on-demand transport. In *2022 3rd International Conference on Computation, Automation and Knowledge Management (ICCAKM)* (pp. 1–6). IEEE.
21. Statista. AI Use Cases in Consumer Goods and Retail Industry Worldwide as of 2020. 2020. Available online: https://www.statista.com/statistics/1197958/ai-use-cases-consumer-goods-retail-global/. (Accessed on 5 December 2021).
22. Anshu, K., Gaur, L., & Singh, G. (2022). Impact of customer experience on attitude and repurchase intention in online grocery retailing: A moderation mechanism of value Co-creation. *Journal of Retailing and Consumer Services, 64*, 102798.
23. Afaq, A., & Gaur, L. (2021, November). The rise of robots to help combat covid-19. In *2021 International Conference on Technological Advancements and Innovations (ICTAI)* (pp. 69–74). IEEE.
24. Khalid, O., Khan, S. U., & Zomaya, A. Y. (Eds.). (2019). *Big data recommender systems: Algorithms, architectures, big data, security and trust* (Vol. 1). Institution of Engineering and Technology.
25. Sorbán, K. (2021). Ethical and legal implications of using AI-powered recommendation systems in streaming services. *Információs Társadalom: Társadalomtudományi Folyóirat, 21*(2), 63–82.

26. Sharma, S., Singh, G., Gaur, L., & Afaq, A. (2022). Exploring customer adoption of autonomous shopping systems. *Telematics and Informatics, 73*, 101861.
27. Lundberg, H., Mowla, N. I., Abedin, S. F., Thar, K., Mahmood, A., Gidlund, M., & Raza, S. (2022). Experimental analysis of trustworthy in-vehicle intrusion detection system using explainable artificial intelligence (XAI). *IEEE Access, 10*, 102831–102841.
28. Cirqueira, D., Nedbal, D., Helfert, M., & Bezbradica, M. (2020). Scenario-based requirements elicitation for user-centric explainable AI: A case in fraud detection. In *Machine Learning and Knowledge Extraction: 4th IFIP TC 5, TC 12, WG 8.4, WG 8.9, WG 12.9 International Cross-Domain Conference, CD-MAKE 2020, Dublin, Ireland, August 25–28, 2020, Proceedings 4* (pp. 321–341). Springer International Publishing.
29. Yang, G., Ye, Q., & Xia, J. (2022). Unbox the black-box for the medical explainable AI via multi-modal and multi-centre data fusion: A mini-review, two showcases and beyond. *Information Fusion, 77*, 29–52.
30. Cirqueira, D., Helfert, M., & Bezbradica, M. (2021, July). Towards design principles for user-centric explainable AI in fraud detection. In *Artificial Intelligence in HCI: Second International Conference, AI-HCI 2021, Held as Part of the 23rd HCI International Conference, HCII 2021, Virtual Event, July 24–29, 2021, Proceedings* (pp. 21–40). Springer International Publishing.
31. Liao, Q. V., & Varshney, K. R. (2021). Human-centered explainable ai (xai): From algorithms to user experiences. *arXiv preprint* arXiv:2110.10790.
32. Rodrigues, V. F., Policarpo, L. M., da Silveira, D. E., da Rosa Righi, R., da Costa, C. A., Barbosa, J. L. V., Antunes, R. S., Scorsatto, R., & Arcot, T. (2022). Fraud detection and prevention in e-commerce: A systematic literature review. *Electronic Commerce Research and Applications*, 101207.
33. Adadi, A., & Berrada, M. (2018). Peeking inside the black-box: A survey on explainable artificial intelligence (XAI). *IEEE access, 6*, 52138–52160.
34. Gaur, L., Sahoo, B.M. (2022). Intelligent transportation technology enablers. In: Explainable Artificial Intelligence for Intelligent Transportation Systems. Springer. https://doi.org/10.1007/978-3-031-09644-0_2
35. Loveleen Gaur, Mansi Ratta, Adesh Gaur (2022), Future of DeepFakes and Ectypes In: Deepfakes. CRC Press, 9781003231493
36. Gaur, L., & Sahoo, B.M. (2022). Explainable AI in ITS: Ethical concerns. In: Explainable artificial intelligence for intelligent transportation systems. Springer. https://doi.org/10.1007/978-3-031-09644-0_5
37. Gaur, L., & Sahoo, B.M. (2022). Intelligent transportation system: Modern business models. In: Explainable Artificial Intelligence for Intelligent Transportation Systems. Springer. https://doi.org/10.1007/978-3-031-09644-0_4.
38. Gaur, L., Jhanjhi, N. Z., Bakshi, S., & Gupta, P. (2022). Analyzing consequences of artificial intelligence on jobs using topic modeling and keyword extraction. In 2022 2nd International Conference on Innovative Practices in Technology and Management (ICIPTM) (pp. 435–440). https://doi.org/10.1109/ICIPTM54933.2022.9754064.
39. Gaur, L., Bhandari, M., Razdan, T., Mallik, S., & Zhao, Z. (2022). Explanation-driven deep learning model for prediction of brain tumour status using MRI image data. *Frontiers in Genetics, 13*, 822666.
40. Ekhart, N. (2022). Taking down malicious webshops: Designing Explainable AI against growing e-commerce fraud.
41. Yılmaz Benk, G., Badur, B., & Mardikyan, S. (2022). A new 360° framework to predict customer lifetime value for multi-category E-commerce companies using a multi-output deep neural network and explainable artificial intelligence. *Information, 13*(8), 373.
42. Chaudhary, M., Gaur, L., & Chakrabarti, A. (2022, November). Detecting the employee satisfaction in retail: A Latent Dirichlet Allocation and Machine Learning approach. In *2022 3rd International Conference on Computation, Automation and Knowledge Management (ICCAKM)* (pp. 1–6). IEEE.
43. Matuszelański, K., & Kopczewska, K. (2022). Customer churn in retail E-commerce business: Spatial and machine learning approach. *Journal of Theoretical and Applied Electronic Commerce Research, 17*(1), 165–198.

44. Waltl, B., & Vogl, R. (2018). Increasing transparency in algorithmic-decision-making with explainable AI. *Datenschutz und Datensicherheit-DuD, 42*(10), 613–617.
45. Hulsen, T. (2023). Explainable Artificial Intelligence (XAI) in Healthcare.
46. Ferreira, J. J., & Monteiro, M. S. (2020). What are people doing about XAI user experience? A survey on AI explainability research and practice. In *Design, User Experience, and Usability. Design for Contemporary Interactive Environments: 9th International Conference, DUXU 2020, Held as Part of the 22nd HCI International Conference, HCII 2020, Copenhagen, Denmark, July 19–24, 2020, Proceedings, Part II 22* (pp. 56–73). Springer International Publishing.
47. Islam, A. S., Ahmed, S., & Khan, R. H. (2022, March). A review on e-commerce system in bangladesh: An empirical study. In *Proceedings of the 2nd International Conference on Computing Advancements* (pp. 269–276).
48. Bradley, C., Wu, D., Tang, H., Singh, I., Wydant, K., Capps, B., Wong, K., Agostinelli, F., Irvi, M., Srivastava, B., & Srivastava, B. (2022, November). Explainable artificial intelligence (XAI) user interface design for solving a Rubik's Cube. In *HCI International 2022–Late Breaking Posters: 24th International Conference on Human-Computer Interaction, HCII 2022, Virtual Event, June 26–July 1, 2022, Proceedings, Part II* (pp. 605–612). Springer Nature Switzerland.
49. Antoniadi, A. M., Du, Y., Guendouz, Y., Wei, L., Mazo, C., Becker, B. A., & Mooney, C. (2021). Current challenges and future opportunities for XAI in machine learning-based clinical decision support systems: A systematic review. *Applied Sciences, 11*(11), 5088.
50. Ravi, M., Negi, A., & Chitnis, S. (2022, April). A comparative review of expert systems, recommender systems, and explainable AI. In *2022 IEEE 7th International conference for Convergence in Technology (I2CT)* (pp. 1–8). IEEE.
51. Ahmed, I., Jeon, G., & Piccialli, F. (2022). From artificial intelligence to explainable artificial intelligence in industry 4.0: A survey on what, how, and where. *IEEE Transactions on Industrial Informatics, 18*(8), 5031–5042.
52. Gaur, L., Afaq, A., Singh, G., & Dwivedi, Y. K. (2021). Role of artificial intelligence and robotics to foster the touchless travel during a pandemic: A review and research agenda. *International Journal of Contemporary Hospitality Management, 33*(11), 4079–4098.
53. Wolofsky, S. (2020). What's Your Privacy Worth on the Global Tech Market? Weighing the Cost of Protecting Consumer Data against the Risk That New Legislation May Stifle Competition and Innovation during this global, technological revolution. *Fordham Int'l LJ, 44*, 1149.
54. Brendel, A. B., Mirbabaie, M., Lembcke, T. B., & Hofeditz, L. (2021). Ethical management of artificial intelligence. *Sustainability, 13*(4), 1974.
55. Suresh, A., & Rani, N. J. (2020). Consumer perception towards artificial intelligence in E-commerce with reference to Chennai city, India. *Journal of Information Technology and Economic Development, 11*(1), 1–14.

Explainable Artificial Intelligence (XAI) for Managing Customer Needs in E-Commerce: A Systematic Review

Koti Tejasvi, V. Lokeshwari Vinya, Jagini Naga Padmaja, Ruqqaiaya Begum, and M. A. Jabbar

Abstract Businesses across industries have changed how they operate as a result of the introduction and adoption of technology. Importantly, significant technological advancements in e-commerce try to persuade consumers to purchase particular goods and brands. AI is increasingly used as a vital new tool for personalization and product customization to meet specific needs. It provides insights into the decision-making criteria, elements, and data required to provide a recommendation. The machine learning field known as XAI studies and strives to understand the models and techniques utilized in the black box decisions produced by AI systems. In order to deploy explainable XAI systems, this study suggested that ML models need to be improved in order to make them easier to comprehend and interpret. A branch of machine learning known as XAI studies and aims to understand the models and processes involved in how AI systems make decisions in a "black box." It offers insights into the considerations, factors, and information needed to generate a suggestion. This study made the recommendation that ML models be enhanced, making them interpretable and understandable, in order to deploy explainable XAI systems. This paper addresses this issue by examining and analyzing recent work in XAI methodologies, needs, principles, applications, and case studies. We introduce a novel XAI approach that facilitates the development of explainable models while maintaining a high level of learning performance.

Keywords Layer-Wise Relevance Propagation (LRP) · Generalized Additive Model (GAM) · SHapley Additive Explanations (SHAP) · Explainable Artificial Intelligence (XAI) · Local Interpretable Model Agnostic Explanations (LIME)

K. Tejasvi (✉) · V. L. Vinya · J. N. Padmaja · R. Begum
Department of Computer Science and Engineering, Vardhaman College of Engineering, Shamshabad, India
e-mail: kotitejasvi@gmail.com

M. A. Jabbar
Department of CSE-AIML, Vardhaman College of Engineering, Hyderabad, India

L. Gaur and A. Abraham (eds.), *Role of Explainable Artificial Intelligence in E-Commerce*, Studies in Computational Intelligence 1094,
https://doi.org/10.1007/978-3-031-55615-9_2

1 Introduction

Technological development continues to open up new prospects for individuals working in a wide range of industries, which benefits the efficacy, efficiency, and cost-effectiveness of the services offered by companies. The most recent advancements in cloud computing block chain, and AI may open up new business prospects for entrepreneurs. Computer systems also influence and enhance client interactions with business organizations. The move towards greater usage of technology has led to the creation of intelligent systems that can manage and monitor business models with less human participation.AI systems that can satisfy customer desires across many industries are necessary for the current economy [1]. AI is crucial for keeping an eye on the business environment, figuring out client needs, and putting in place the appropriate strategies with little to no human involvement.

Therefore, it bridges the gap between consumer needs and efficient or high-quality services. As a result, AI is changing the economic landscape and bringing about adjustments that can greatly benefit customers and business owners. It is becoming more and more common in companies, particularly in marketing, financial management, and business administration. The opportunities that AI produces lead to significant changes in the world's economic structures. E-commerce gains the most when AI is used more frequently to improve service effectiveness and quality [2]. The potential harm caused by human error is reduced with the help of AI. As a result, businesses will still benefit substantially from AI even though it may lead to fewer employment opportunities. Notably, e-commerce has flourished and grown because of powerful forces like AI. In e-commerce, artificial intelligence (AI) technologies enable network marketing, electronic payments, and control of the logistics required to make products available to consumers. For e-commerce food enterprises, AI is crucial since it maintains hygienic conditions at the manufacturing site and ensures the creation of safe food. It also helps maintain the cooking equipment as clean as possible. Automated systems collect, analyses, and review data more quickly than humans do. E-commerce is aided by AI in identifying market trends and clients' changing needs. When performing certain duties in e-commerce, such as forecasting demand and supply chain mechanisms, human intelligence may frequently be constrained. To handle the growing challenges in e-commerce, AI extends and simulates human intelligence [1]. Soni, for instance, found that AI aids e-commerce sites in managing and keeping track of their customers.

To ensure that customers receive high-quality services, a business might utilize AI to gather a range of data and evaluate them. This enables e-commerce platforms to better understand the factors influencing the purchasing decisions of its current and potential clients. It improves messaging and chat app interactions between e-commerce companies and their customers. E-commerce companies employ automation strategies to cut down on operational redundancy. Automated responses to consumer inquiries are made possible by AI. Threats and obstacles to e-commerce and customers, however, put a cap on AI's effectiveness and efficiency in fulfilling business expectations. In view of changing consumer needs, it is crucial to look into e-commerce potential and constraints.

To allow trust, XAI is primarily required. To be able to believe the model's conclusions, it is crucial to explain them. Particularly when a prediction-based choice is made and the implications for human safety are involved. To recognize and comprehend the bias in these choices is another reason why XAI is required. It is not surprising that bias can exist in the datasets used in practice, even the ones that are well established, since it can be observed in many facets of everyday life [3]. Unfairness between discriminatory characteristics like gender or ethnicity can be caused by bias issues. The latest Netflix documentary that focuses on the significance of unfair machine learning algorithms and their impact on society also investigates this issue. We can identify and comprehend fairness problems with the aid of XAI, which will enable us to get rid of them. If you still need persuading, there are a plenty more justification listed in the why of explainable AI.

- Four explainable artificial intelligence principles are outlined by the National Institute of Standards (NIST), a subsidiary of the U.S. Department of Commerce, to better define what XAI is [4]: "Evidence, support, or reasoning for each output" should be provided by an AI system.
- An AI system ought to give its customers explanations they can follow.
- An explanation should accurately reflect the AI system's process.
- AI should operate only under the conditions it was designed for and not provide output when it lacks confidence.

2 XAI Methods/Techniques

See Fig. 1.

2.1 Layer-Wise Relevance Propagation (LRP)

The internals of the model, including the topology, weights, and activations, must be accessible for Layer-wise Relevance Propagation (LRP), a propagation-based explanation technique. However, LRP is able to simplify the model as a result of this extra information, which helps to solve the explanation issue more effectively [5]. However, LRP is able to simplify the model as a result of this extra information, which helps to solve the explanation issue more effectively. Instead of explaining the prediction of a deep neural network in one step as model agnostic methods (e.g., pixels) would, LRP uses the network structure to redistribute the explanatory factors (known as relevance R) layer by layer, starting from the model's output, onto the input variables. Due to the fact that each redistribution only affects two neighbouring layers, it may be said that it resolves a simple explanation problem (see interpretation of LRP as Deep Taylor Decomposition).LRP is a well-liked explanation technique that has

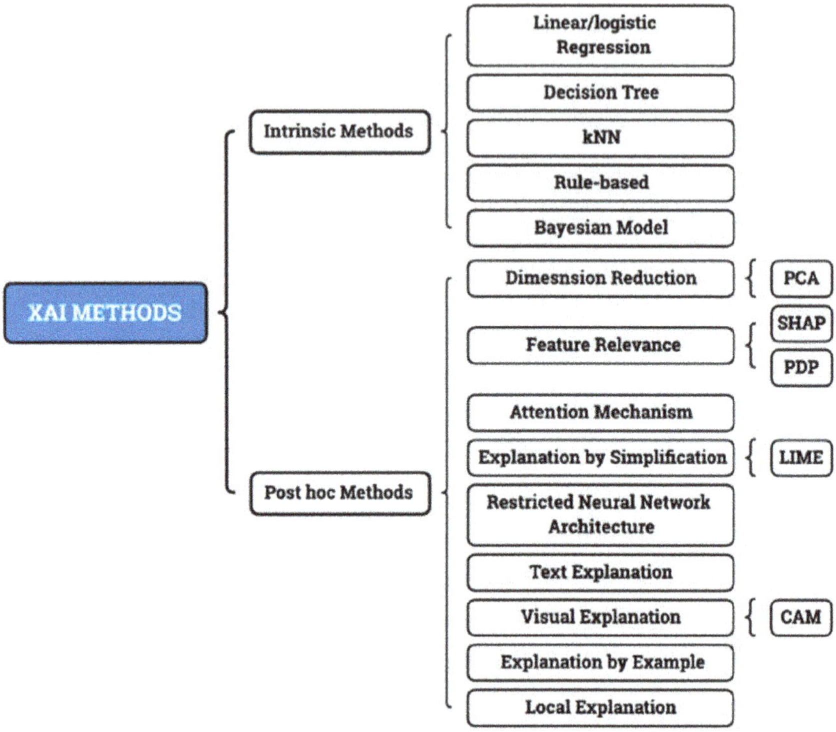

Fig. 1 Overview of methods/techniques in XAI

been used in a variety of fields, such as computer vision, natural language processing, EEG analysis, and weather, among others. High computational efficiency, theoretical support that makes it reliable, a robust explanation technique, a long history, and high popularity are the primary benefits of LRP. Restricted flexibility is the cost of the benefits; Adapting the current redistribution rules carefully may be necessary in some cases, such as when using unique model designs.

LRP computes relevance values for the input variables as well as the weights, biases, and individual neurons of the neural network in order to appropriately trim and quantize the model. The concept is straightforward: by simply removing the irrelevant components from the neural network, we can increase coding efficiency and accelerate processing because LRP explanations inform us which neural network components are pertinent.

2.2 *Local Interpretable Model Agnostic Explanations (LIME)*

Due to the lack of knowledge about the model's internals, such as the topology, learning parameters (weights, biases), and activation values in the case of neural networks, LIME methods are very flexible and generally applicable. LIME's main objective is to make the [Math Processing Error] prediction of a complex model more understandable. The surrogate-based explanation approach is another name for LIME. It is a simple to understand surrogate paradigm [6]. As a result, LIME explains the forecasts of a surrogate model instead of the target model itself, which is a local approximation of the target model. Although extensive, LIME's central concept is actually quite clear-cut and straightforward. Let's look at the word itself and what it means:

- *Model agnosticism*: LIME has the ability to explain any particular supervised learning model by treating each one as a unique "black-box" because to its model agnosticism trait. This implies that LIME can accommodate practically any paradigm that is currently in use.
- *Local explanations*: LIME provides explanations that are locally accurate within the vicinity or surroundings of the data or sample being explained.

LIME has been used in a variety of application domains, proving the usefulness of this technique regardless of the model. It is apparent that LIME's use of a surrogate model only serves to partially resolve the explanation issue. As a result, the surrogate fit's quality—which may call for dense sampling and consequently raise computational costs—largely determines how well the theory fits the data.

2.3 *Counterfactual Method*

Utilizing the counterfactual impact evaluation approach, ascertain the proportion of the genuine improvement that has been observed that can be attributable to the intervention's impacts (such as a rise in income). (Because the intervention might not be the only reason for the improvement; other variables, such the expansion of the economy as a whole, could also be to blame. Counterfactual arguments attempt to justify the model using the following short and concise claim:

> Y would not have happened if X hadn't happened.

Here, we can consider X to be a data feature and Y to be the result or a predicted number for a specific instance. Counterfactuals do not necessarily have to be novel combinations of feature values, unlike prototypes, a different technique under the category of Example Based Explanations.

Numerous studies have been conducted on counterfactuals across a variety of disciplines, particularly in logic, statistics, and cognitive science. For XAI, hypothetical fall under the category of example-based methods. They are founded on methods that figure out what adjustments should be made to the instance data point in order to alter its prediction to the desired result. The characteristics of an effective counterfactual are proximity, plausibility, sparsity, diversity, and feasibility. Six different categories that collectively reflect the "master theoretical algorithm" from which each algorithm was descended. These classes include methodologies that are (1) instance-centric, (2) constraint-centric, (3) genetic-centric, (4) regression-centric, (5) game theory-centric, (6) case-based reasoning-centric, and (7) network-centric.

2.4 SHapley Additive Explanations (SHAP)

A well-known XAI system called SHapley Additive explanation (SHAP) can offer model-independent local explain ability for datasets that are tabular, visual, and text-based. Shapley values are the foundation of SHAP; this idea is frequently employed in game theory. The SHAP method of explanation uses SHapley values from coalitional game theory to equitably distribute the gain among players when contributions from participants are not equal. Shapley values are an idea from game theory and economics that describe a way to fairly distribute a game's prize money among a group of players [6].

In cooperative games, SHAP makes advantage of four fairness features to distribute prizes equally among players: 1. Dummy, which states that a feature that has no bearing on the game must not be used. 2. Additivity, which states that amounts must add up to the overall game result. 3. Symmetry, which states that a player who contributes more to the game cannot receive a smaller reward. 4. Efficiency, which states that the prediction must fairly attribute the feature values.

2.5 Generalized Additive Model (GAM)

In statistics, a GAM is a generalized linear model in which the linear predictor depends linearly on some predictor variables' unknown smooth functions, with the emphasis being on inferring these smooth functions [6].Trevor Hastie and Robert Tibshirani created GAMs in the beginning to combine additive models' characteristics with those of generalized linear models. The model links some predictor factors, xi, to a univariate response variable, Y. An exponential family distribution for Y (such as the normal, binomial, or Poisson distributions) and a link function g (such as the identity or log functions) are created in order to connect the expected value of Y to the predictor variables.

$$g(E(Y)) = \beta_0 + f_1(x_1) + f_2(x_2) + \cdots + f_m(x_m).$$

3 Applications of Explainable AI

Healthcare: One of the research areas of XAI with the most activity is healthcare. Clinical decision-making, disease diagnosis, and recommendation-based healthcare are all topics of research in this area. In the healthcare industry, doctors with little technological expertise are the main consumers of AI-based technologies. In addition, they frequently hold their own viewpoints in regard to diagnosing diseases and making therapeutic decisions. Hence, the explanation requested by doctors should have enough visual and written information as well as pertinent contextual references. A communicative and interactive user interface should be included in healthcare-based products like fitness apps and dietary advice [7].

Media and entertainment: Systems for recommending music, movies, works of art for museums and websites, news articles, and arcade gaming systems are all included in this research field. Both music and movie recommendation users want personalized suggestions delivered in a variety of explanation formats, and they require explanations that include details about the personalized recommendations as well as details about the personal data utilized [8]. The volume of information being given also raises concerns from users, who worry that it may lead to cognitive overload.

Education: Intelligent tutoring systems, systems that help universities decide which students to admit, and systems that estimate grades are all included in the education area. Explanations increase the usability of the system, according to studies on intelligent tutoring systems [9]. The justifications for certain decision-making processes, such as the admissions decision-making process, should also be included in the explanations, along with details on the behaviour and operation of the system. Users frequently claim that humans, not a machine that just uses algorithms, should determine admission decisions.

Transportation: The transportation category also includes flight rerouting systems for the aviation industry as well as navigational systems and autonomous vehicle decision support systems. Domain experts will benefit most from case-based explanations in autonomous vehicle decision support systems [4]. Case-based explanations connect the rationale behind a choice to examples of actual usage. The user may receive explanations (hints) in text, light signals, or a visual format, among other formats. As a result of users' requirement for prompt justifications and explanations for each choice, the user criteria for navigation systems are slightly different. An analogous pattern of behaviour can be seen in flight rerouting systems.

Finance: Financial use cases of XAI research include insurance, financial fraud detection, and loan applications. The operator should be able to examine the borrower's details, credit history, and other demographic data. Additionally, it was suggested that when creating an explainable system, the designers needed to comprehend and relate to the user's mental model. This is crucial for financial operations including loan approvals and insurance claims. An effective XAI system should be able to identify the incorrect mental model and modify itself accordingly [10].

E-commerce: The authors have researched the social transparency and design principles that promote confidence in the judgement of AI-based systems used in e-commerce. The promotion of social transparency calls for a deeper comprehension of artificial intelligence-based systems. XAI transparency can be used as a helpful marketing strategy, even though its long-term effects have not been studied. A less visible algorithm may also exacerbate privacy issues because analytics for internet advertising leverages personal data [11]. An explainable AI-based solution that combines text and augmented reality with online buying is better. Customers are more trustworthy, which makes for a better online shopping experience.

Human resource management: Many factors affect how willing employees are to accept decisions made by artificial intelligence in human resource management. Many employees believe that making decisions might be biased, dishonest, and an invasion of privacy. As a result, adopting predictions based on AI may be mentally taxing. Closing the knowledge gap and enhancing transparency and interpretability would help us understand decision-making better. The possibility that a system will be adopted and the attitude of the user towards the system may both be improved by working collaboratively with human users throughout the design phase [12]. The algorithmic hiring method is gaining popularity. It's becoming more and more common to use an algorithm for hiring. Explaining the decision-making process, describing the candidate's evaluation results, and demonstrating comparable recruitments within the company are all examples of the explain ability criteria for recruiters. Using information from a past, comparable recruitment, the recruiters may be able to see potential bias in the decision-making process. Furthermore, it would be a good idea to provide a summary of previous work each time a recruiter logs into the system if they are using various recruiters and rotating shifts.

Digital Assistants: According to earlier research, users are more likely to trust virtual assistants when they are more engaged and human-like. Users find encouragement and attraction in verbal comments, gestures, and voice, particularly those that refer to phonemes [13]. An XAI system must also provide linguistic justifications for end users. As a consequence, an interactive agent with a well-balanced combination of explicable AI approaches and a proper linguistics representation can improve a system's dependability and user-centeredness.

E-Governance: In order to determine how fair machine-learning algorithms are seen by users and whether these algorithms require explanations, empirical evaluations of a criminal justice use case were conducted. To improve the system's comprehension, credibility, and trustworthiness, the system should include explanations of how the algorithm functions, the variables that affect decision-making, and the accessibility of contextual data. Similar to how investigations into the use of algorithms in immigration services demonstrate that not all decisions must be made using algorithms, despite the fact that they can support decision-making [14]. The white box approach, which employs explainable AI, has been shown to enhance judgement, according to one study. Therefore, e-governance need human input when making important choices.

Social Networking: Users desire both "why" and "why not" explanations for certain system operations, according to social networking studies. Thus, developers may rapidly offer user log data, data linked to mental models, and contextual data [15]. Additionally, a successful explainable AI system requires human user involvement throughout the design process via a particular communication channel.

4 Case Study for Compliance and Legal Professionals: XAI in Consumer Lending

The ML team of the business unit, which is made up of over 12 researchers, data scientists, and engineers, deployed a binary classification ML model as the subject of the case study [15]. Customers request credit lines, and the model decides whether to approve or reject the request. The internal, non-technical audience in this scenario was mostly composed of compliance and legal experts, and the ML team used both technical and non-technical XAI methodologies to explain the model's ethical and legal consequences to this audience [16]. The case study, which focuses on a binary classification ML model that was deployed by the unit, is about the unit, which has a dedicated ML team of more than 12 researchers, data scientists, and engineers. Customers request credit lines, and the model decides whether to approve or reject the request.

4.1 Target Audiences and AI Explanation Expectations

Together with the Compliance team and the Law team, the ML team developed AI explanations. Their top aims and objectives are shown in Table 1 [15]. The members of both teams all hold bachelor's degrees or higher in business, finance, or the liberal arts; however, none of them have degrees in science, technology, engineering, or mathematics, nor do they have any prior experience in these sectors. All but two have advanced degrees in law. Each of them could appropriately (1) explain statistical outliers and provide instances, (2) depict a "bell curve," and (3) conceptually comprehend what the standard deviation indicates.The mathematical expressions and statistical limitations of those phrases, however, are unfamiliar to them.This data led us to the conclusion that these teams all fell under the umbrella of what we now consider to be a non-technical. Following is how we determined the compliance and legal team's request for AI justifications:

- The explanations should be statistically correct while remaining basic enough for people to grasp.
- Decisions (like overturning a refusal decision) or opinions (like determining if the model conforms with legal standards) will be made using the explanations.
- To support the explanations, supporting information such as documentation, precise quantitative data, and back testing would be required.

Table 1 Stakeholder teams in the deployment of a new machine learning model

Team	Top priority	Goal for XAI
Machine learning	When compared to existing underwriting requirements, the model generates higher approval rates and lower default rates	To debug, fine-tune, and repair model features and behaviors
Compliance	Model conforms with all existing consumer	Evidence that the model is not biased towards protected groups
	Lending regulations, including those governing fair	
	Lending and equitable credit	
Legal	Every eventuality that the law includes or implies should be covered by the model	To establish whether he model is sufficiently complete in terms of legal considerations

4.2 *Tools and Procedures*

Compliance and Legal received documentation and model explanations for every development iteration. These records were those that were used the most frequently among them:

- A description of diverging impact (DI) for protected groups (see Table 2 for an example).
- A list of the model's distinguishing features
- Individual conditional expectation (ICE) plots, partial dependency plots, decision tree surrogates, Shapley explanations, and other simplified XAI techniques.
- A collection of the model features and their relative weights
    - An additional list of "corner cases" from the back test would be given to the Compliance and Legal teams as well. These "corner cases" include both applications that had decisions reversed by a manual credit review team and applications that were chosen at random.

The ML team used the well-known ratio representation of approval ratings across varied groups to build the DI and assess the model's fairness. Because loan functions are not permitted to collect protected group information (such as race or gender), the

Table 2 To the compliance and legal departments, an example summary of differential impact was presented

Groups	Approval	Total	Rate	Impact ratio
Baseline	88,218.48	1,256,840.72	0.08	1.02
IPG 1	94,071.22	1,495,032.24	0.05	0.89
IPG 2	23,952.79	288,583.94	0.008	1.15

The numbers are made up, and the decimal point is rounded to two digits. IPG stands for inferred protected group

ML team went to alternative data sources to assess protected group status, resulting in the non-integer approval and total numbers in Table 2 [15].

If compliance and legal determined that particular parts posed regulatory risk or that any protected group had been harmed, the ML team would alter the model. The ML team employed sophisticated XAI approaches to troubleshoot and resolve the problem. For example, if a global tree surrogate of low error (e.g., RMSE, MAPE) revealed an unusual split for a subpopulation, the team may consider using tree SHAP (SHapley Additive explanations) to obtain locally accurate feature contributions and to generate the adverse action codes required by regulations. Once the "corrected" model had passed ML team testing and a reduced version of the technical XAI procedures had been built for them, the review and remedy cycle with Compliance and Legal would be restarted.

4.3 Learning Experiences

Some of the most essential lessons we learned from the case study.

- **Know the audience**: This is known as "user modeling," a more formal word for a well-established research method in human–computer interfaces (HCI) [17]. Provide sufficient contexts for researchers to develop techniques and perform studies by providing in-depth descriptions of the desirable characteristics of the targeted users, such as users' motives, prior knowledge, behavioral patterns, etc. For the effective deployment of XAI in our case study, it was essential to comprehend the priorities and interests of the compliance and legal teams.
- **Stakeholders gained knowledge to construct XAI mental models**: "Mental models" are internal representations of a scenario that have the capacity to anticipate and explain occurrences to their own bearers, as is widely known in psychology and HCI [18]. The comprehension and application of AI explanations would be improved if non-technical audiences were assisted in developing appropriate expectations about XAI and how XAI may or may not assist with their ultimate goals.
- **Customize your communications**: The results of XAI that AI specialists can grasp may not be clear to audiences that are not technically sophisticated. Giving explanations in a way that non-technical audiences can easily absorb with their own prior knowledge and topic expertise would be a big step towards bridging the gap between technical and non-technical XAI.
- **Include social science knowledge and expertise**: The dissemination of findings to audiences that are typically educated but not necessarily professionals using comprehensible forms has a long history in both psychology and economics. For instance, the "disparate impact" for various groups, which the ML team supplied to the compliance team, has received extensive research in psychology and economics [19] and is regarded by working professionals as a norm for fairness. They are actually easier to understand by non-technical audiences than

the fairness criteria applied by AI experts. Similar to this, domain knowledge is combined to make XAI more approachable for a highly educated but primarily non-AI expert audience, for example by building XAI procedures in the medical domains [20].

5 Modern XAI Technologies

Figure 2a [2] summarizes the present Modern XAI Technologies and shows its high-level structure (part a). On the one hand, attention has been made on the order, dependence, and current sequence of research activities, which are frequently carried out by multiple scholars. The majority of the time, this process starts with the entry of data, which is then utilized for modelling using connectionist data-driven learning or knowledge-based paradigms based on symbolic reasoning. After a model is developed, it is analyzed using XAI techniques in order to uncover additional data and support the model's interpretability. The end users of the models are provided with one or more explanations in this step so that they can understand the models. Finally, only a small number of scholars have offered strategies for evaluating this layer of explainability, either by establishing formal, objective standards or by including human-centered evaluation with model developers and end users. The optimum structure for XAI, on the other hand, is shown in Fig. 2b [2], in our opinion (part b). Since explanators are what end users will ultimately interact with, they should be the primary focus of this situation. The various characteristics related to the psychological construct of explain ability should be taken into consideration when developing explanators. Academics can then focus on the modelling step, ideally using both the connectionist and symbolic AI paradigms. This will make it feasible to design models that are fundamentally interpretable and durable in terms of accuracy during all phases of development. The third and final phase should focus on developing an interactive interface for model interpretability and inference explain ability evaluation using a human-in-the-loop method.

6 Conclusion and Future Challenges

Beyond what has been accomplished so far in the XAI area, we have broadened the scope of our discussions and are now concentrating on the concept of responsible AI, a paradigm that imposes several AI principles, such as fairness, transparency, and privacy, to be respected when putting AI models into practice. Fairness implications of XAI have also been carefully considered. Figure 3 [21] illustrates the idea of XAI as a vital concept to ensure the guidelines for Responsible AI.

Our opinions on the future of XAI, as voiced in the discussions held throughout this study, agree that a correct understanding of the potential benefits and dangers of XAI approaches is urgently required. We think that needs and constraints for model

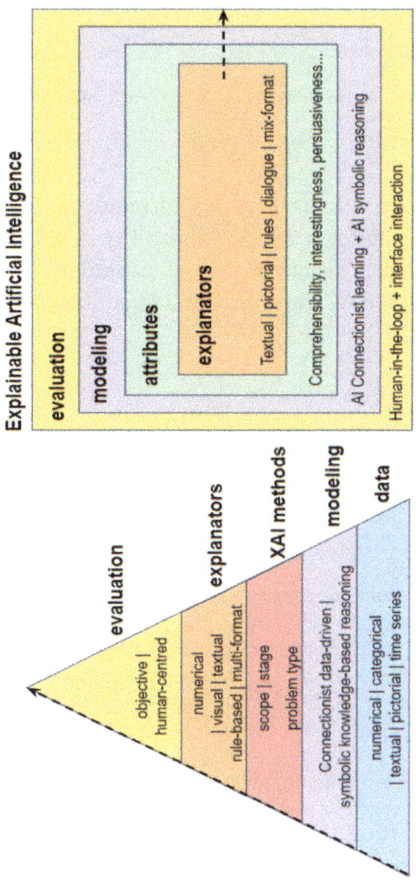

Fig. 2 Modern XAI technologies **(a)** and envisioned **(b)** frameworks for explainable artificial intelligence

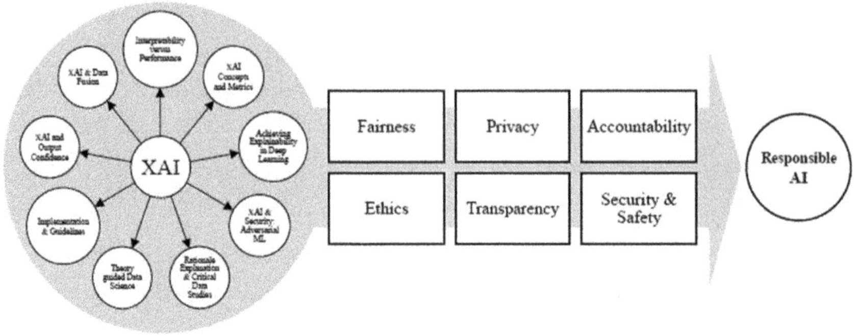

Fig. 3 A summary of the XAI issues covered in this review and their implications for the Responsible AI tenets

interpretability must be addressed concurrently with requirements and constraints for data privacy, model secrecy, fairness, and accountability. A responsible deployment and utilization of AI techniques in enterprises and institutions throughout the world can only be ensured if all three AI tenets are considered jointly.

References

1. Khrais, L. T. (2020). Role of artificial intelligence in shaping consumer demand in E-commerce. *Future Internet, 12*, 226. https://doi.org/10.3390/fi12120226
2. Vilone, G., & Longo, L. (2020, Oct 12). Explainable artificial intelligence: A systematic review. arXiv:2006.00093v4 [cs.AI].
3. Arikan, Z. B. An introduction to explainable AI. https://www.mobiquity.com/insights/an-introduction-to-explainable-artifical-intelligence.
4. Netapp, Mike McNamara**,** netapp.com/blog/explainable-ai/#sub2–1.
5. Zhang, Y., Weng, Y., & Lund, J. (2022). Applications of explainable artificial intelligence in diagnosis and surgery. *Diagnostics, 12*, 237. https://doi.org/10.3390/diagnostics12020237
6. Durmus, M. Inside the black box: 5 Methods for XAI, https://www.aisoma.de/5-methods-for-explainable-ai-xai/.
7. Haque, A. K. M. B., Islam, A. K. M. N., Mikale, P. Explainable artificial intelligence (XAI) from a user perspective: A synthesis of prior literature and problematizing avenues for future research. https://doi.org/10.1016/j.techfore.2022.122120.
8. Ngo, T., Kunkel, J., Ziegler, J. E. Exploring mental models for transparent and controllable recommender systems: A qualitative study. https://doi.org/10.1145/3340631.3394841.
9. Cheng, H., Wang. R., Zhang, Z., O'Connell, F., Gray, T., Harper, F. M., & Zhu, H. Explaining decision-making algorithms through UI: Strategies to help non-expert stakeholders. https://doi.org/10.1145/3290605.3300789.
10. Schneider, T., Ghellal, S., Love, S., Gerlicher, A. R. S. Increasing the user experience in autonomous driving through different feedback modalities. https://doi.org/10.1145/3397481.3450687.
11. Ehsan, U., Liao, Q. V., Muller, M., Weisz, J. D. Expanding explainability: Towards social transparency in AI systems. https://doi.org/10.1145/3411764.3445188.

12. Binns, R. D., Van-Kleek, M.G., Veale, M., Lyngs, U., Zhao, J., Nigel, R., & Shadbolt. It's reducing a human being to a percentage: Perceptions of justice in algorithmic decisions. https://doi.org/10.1145/3173574.3173951.
13. Weitz, K., Schiller, D., Schlagowski, R., Huber, T. B., & Andrè, E. Do you trust me?": Increasing user-trust by integrating virtual agents in explainable AI interaction design. https://doi.org/10.1145/3308532.3329441.
14. Janssen, M., Hartog, M., & Kuk, G. Will algorithms blind people? The effect of explainable AI and decision-makers' experience on AI-supported decision-making in government. https://doi.org/10.1177/0894439320980118.
15. Jiang, H., Senge, E. (2021, Dec 2). On two XAI cultures: A case study of non-technical explanations in deployed AI system. arXiv:2112.01016v1 [cs. HC].
16. Yin, M., Vaughan, J. W. W., & Wallach, H. M. Understanding the effect of accuracy on trust in machine learning models. https://doi.org/10.1145/3290605.3300509.
17. Fischer, G. (2001). User modeling in human–computer interaction. *User Modeling and User-adapted Interaction, 11*(1–2), 65–86.
18. Johnson-Laird, P. N. (1983). Mental models: Towards a cognitive science of language, inference, and consciousness. No. 6. Harvard University Press.
19. Avery, R. B., Brevoort, K. P., & Canner, G. (2012). Does credit scoring produce a disparate impact? *Real Estate Economics, 40*, S65–S114.
20. Ahmad, M. A., Eckert, C., & Teredesai, A. (2018). Interpretable machine learning in healthcare. In *Proceedings of the 2018 ACM International Conference on Bioinformatics, Computational Biology, and Health Informatics* (pp. 559–560).
21. Arrieta, A. B., D´ıaz-Rodr´ıguez, N., Ser, J. D., Bennetot, A., Tabik, S., Barbado, A., Garcia, S., Gil-Lopez, S., Molina, D., Benjamin, R., Chatila, R., & Herrera, F. 92019, Dec 26). Explainable artificial intelligence (XAI): Concepts, taxonomies, opportunities and challenges toward responsible AI arXiv:1910.10045v2 [cs.AI].

Decoding the Recommender System: A Comprehensive Guide to Explainable AI in E-commerce

Garima Sahu and Loveleen Gaur

Abstract The rapid growth of e-commerce has resulted in an increasingly competitive landscape where businesses strive to provide personalized and engaging experiences to their customers. Recommender systems, powered by advanced algorithms and artificial intelligence, are central to this effort, curating tailored suggestions for products, services, and content. However, the complex and opaque decision-making processes of these systems often act as black boxes, limiting user understanding and trust. This chapter delves into the exclusive roles of explainable AI in the decision-making processes of recommender systems within the context of e-commerce, highlighting its importance in fostering trustworthiness, ensuring ethical and legal compliance, and facilitating debugging and model improvement. We explore various types of explanations, techniques for generating explanations, and real-world examples of explainable recommender systems. In conclusion, explainable AI is an indispensable component of recommender systems, playing a critical role in enhancing user trust and engagement, ultimately leading to improved customer satisfaction and increased revenues for e-commerce businesses. As AI systems continue to evolve and become more integrated into our lives, explainability will remain a crucial aspect of their design and implementation.

Keywords E-commerce · Recommender systems · Explainable AI · Trustworthiness · Customer engagement · Customer satisfaction

---

G. Sahu (✉)
Amity International Business School, Amity University, Noida, India
e-mail: garima.sahu18@rediffmail.com

L. Gaur
Taylor's University, Subang Jaya, Malaysia

The University of the South Pacific, Suva, Fiji Islands

L. Gaur and A. Abraham (eds.), *Role of Explainable Artificial Intelligence in E-Commerce*, Studies in Computational Intelligence 1094,
https://doi.org/10.1007/978-3-031-55615-9_3

1 Introduction

E-commerce has grown exponentially over the last few years, transforming the way consumers shop and interact with businesses [49, 50]. As the landscape becomes more competitive, companies are striving to provide a personalized and engaging experience for their customers. Central to this effort is the adoption of recommender systems that curate personalized suggestions for products, services, and content [40, 41]. These systems leverage advanced algorithms and AI techniques to analyze user preferences and behavior, helping businesses optimize their offerings and maximize customer satisfaction [38].

However, many recommender systems operate as black boxes, with complex and opaque decision-making processes that are difficult for users to understand [10]. This lack of transparency can lead to a disconnect between users and the recommendations they receive, reducing trust in the system and hampering its effectiveness [1, 2]. As AI becomes increasingly integrated into our daily lives, there is a growing demand for systems that are not only intelligent but also explainable, interpretable, and accountable [3, 4]. XAI, an emerging field of research and development that aims to make AI systems more transparent and understandable to humans [5–8]. By providing clear explanations for the decisions and actions taken by AI models, XAI seeks to bridge the gap between complex algorithms and human comprehension [38]. This chapter explores the exclusive roles of explainable AI in the decision-making processes of recommender systems within the context of e-commerce.

Understanding the rationale behind AI-generated recommendations is crucial for fostering trust and user confidence. When customers are provided with clear explanations for the suggestions they receive, they are more likely to engage with the recommended products or services, leading to higher conversion rates and increased customer satisfaction [9, 10]. Moreover, explainable AI can help businesses comply with ethical guidelines and legal regulations that emphasize transparency, fairness, and accountability [11]. In addition to enhancing trust and user confidence, explainable AI can also facilitate the debugging and improvement of recommender systems. By making the decision-making processes more transparent, developers and domain experts can identify potential biases, errors, or areas of improvement in the system [50]. This feedback loop is essential for optimizing the performance of the recommender system and ensuring that it delivers relevant and accurate recommendations to users [1–4, 10].

This chapter delves into the types of explanations that can be provided by recommender systems, including content-based explanations, collaborative-based explanations, and hybrid explanations. Each of these approaches offers unique insights into the rationale behind recommendations, and understanding their nuances is crucial for designing effective and transparent recommender systems. And then explores various techniques for generating explanations, such as model-agnostic methods and model-specific methods. Model-agnostic techniques, such as Local Interpretable Model-agnostic Explanations (LIME) and Shapley Additive Explanations (SHAP)

[40], can provide insights into the recommendations without requiring a deep understanding of the underlying model. In contrast, model-specific methods are tailored to specific recommendation models and can offer more fine-grained insights into the decision-making processes.

Furthermore, this chapter also provides real-world examples of explainable recommender systems in e-commerce, such as Amazon and Netflix. These case studies showcase how leading companies are leveraging explainable AI to provide their users with transparent and meaningful recommendations, enhancing user satisfaction and trust in their platforms. Overall, this chapter highlights the critical role of explainable AI in the decision-making processes of recommender systems in the e-commerce domain. By providing users with clear and understandable explanations for the recommendations they receive, businesses can foster trust, user confidence, and engagement, ultimately leading to improved customer satisfaction and increased revenues [12–14]. As AI systems continue to evolve and become more integrated into our lives, explainability will remain an indispensable aspect of their design and implementation.

2 The Need for Explainable AI in Recommender Systems

The integration of explainable AI into recommender systems has become essential for multiple reasons. This section delves deeper into the primary motivations driving the adoption of explainable AI, focusing on enhancing trustworthiness and user confidence, ensuring ethical and legal compliance, and facilitating debugging and model improvement.

2.1 Trustworthiness and User Confidence

Trustworthiness and user confidence are critical factors in determining the success of a recommender system in the e-commerce domain. In an increasingly competitive market, consumers face a plethora of choices, making it difficult for them to decide on which products or services to choose [15]. Recommender systems help users navigate this complex landscape by providing personalized recommendations tailored to their preferences and behaviors [16]. However, if users cannot trust these recommendations or do not understand the rationale behind them, they may feel overwhelmed and disengage from the platform.

Explainable AI plays a vital role in fostering trustworthiness and user confidence in recommender systems. It does so by making the decision-making processes of these systems more transparent, providing users with clear and understandable explanations for the recommendations they receive [17]. By offering insights into the reasons behind a recommendation, explainable AI empowers users to make more informed decisions, increasing their confidence in the recommender system and the e-commerce platform.

There are several ways in which explainable AI can contribute to the trustworthiness and user confidence in recommender systems:

- *Personalized explanations:* Explainable AI can provide users with personalized explanations that are tailored to their preferences and browsing history. By understanding the factors that influence a recommendation, users can better evaluate its relevance and usefulness.
- *Context-aware explanations*: Users often interact with e-commerce platforms in various contexts, such as searching for specific items, browsing for inspiration, or comparing products. Explainable AI can generate context-aware explanations that take these different scenarios into account, offering users more meaningful and relevant insights.
- *Consistency and coherence*: For users to trust a recommender system, the explanations it provides must be consistent and coherent across different recommendations and contexts. Explainable AI can help ensure that the explanations are logical and accurate, preventing confusion and enhancing user confidence.
- *Explainability metrics*: By incorporating quantitative metrics that assess the quality and effectiveness of explanations, explainable AI can provide insights into the system's performance. This allows businesses to optimize their recommender systems and offer explanations that effectively build trust and confidence among users.
- *Feedback mechanisms*: Integrating feedback mechanisms into the recommender system can help improve the quality of explanations over time. Users can provide feedback on the relevance and usefulness of explanations, which can be used to refine and adapt the system's decision-making processes.

Ultimately, the enhanced trustworthiness and user confidence resulting from explainable AI led to increased user engagement, higher conversion rates, and improved customer satisfaction. These factors are crucial for the success of e-commerce platforms, making explainable AI an indispensable component of recommender systems.

2.2 Ethical and Legal Compliance

As AI systems gain increasing autonomy and influence over users' lives, concerns about fairness, accountability, and transparency have come to the forefront. Implementing explainable AI in recommender systems helps address these concerns by providing clear explanations of the decision-making processes, ensuring that the systems operate in an ethical and legally compliant manner [18]. Ethical concerns in recommender systems primarily involve fairness and biases. AI-driven recommendation algorithms may unintentionally perpetuate or exacerbate biases present in the data they process [19]. For instance, if a recommender system is trained on data that predominantly features products purchased by a specific demographic group, the recommendations it generates may be biased towards that group's preferences, disadvantaging other users. Such biases can lead to unfair treatment of users, perpetuating existing inequalities and harming the overall user experience.

Explainable AI can help identify and mitigate these issues by offering insights into the factors contributing to a recommendation. By understanding how the system processes and weighs various features, developers can detect potential biases and implement corrective measures to ensure fairness in recommendations [20, 21]. For example, they may modify the algorithm to give more weight to underrepresented groups or employ techniques such as re-sampling or re-weighting to balance the training data. Providing users with transparent explanations of the factors influencing recommendations also helps raise awareness of potential biases and allows users to make more informed decisions based on these insights. From a legal standpoint, several jurisdictions have introduced regulations emphasizing the importance of transparency and accountability in AI systems. These regulations seek to ensure that AI technologies are used responsibly and that individuals affected by AI-driven decisions are treated fairly and with respect [22].

For instance, the European Union's General Data Protection Regulation (GDPR) includes provisions that grant users the right to explanations for algorithmic decisions that significantly impact them. Article 22 of the GDPR stipulates that individuals have the right not to be subject to a decision based solely on automated processing, including profiling, which produces legal effects concerning them or similarly significantly affects them. Furthermore, Recital 71 of the GDPR states that individuals should have the right to obtain an explanation of the decision reached after such processing [23–25]. By incorporating explainable AI into recommender systems, businesses can adhere to these legal requirements and avoid potential fines and reputational damage. Providing clear and comprehensible explanations of the rationale behind recommendations allows users to understand the factors influencing the system's decisions, ensuring that their rights are respected and upheld [26]. Moreover, transparency in the decision-making process can help businesses demonstrate their commitment to responsible AI practices, fostering trust among users and regulatory authorities alike.

In addition to the GDPR, other legal frameworks and guidelines are being developed to address the ethical and legal challenges posed by AI technologies. For example, the European Commission's Ethics Guidelines for Trustworthy AI outline seven key requirements for AI systems, including transparency, accountability, and fairness. By embracing explainable AI principles, businesses can ensure that their recommender systems align with these guidelines and contribute to the development of a trustworthy AI ecosystem [27]. In summary, explainable AI plays a crucial role in ensuring ethical and legal compliance for recommender systems in e-commerce. By offering transparent explanations of the decision-making processes and addressing potential biases, businesses can demonstrate their commitment to responsible AI practices and foster trust among users and regulators. This, in turn, helps create a more equitable and legally compliant environment in which AI-driven recommender systems can thrive.

2.3 Debugging and Model Improvement

Explainable AI offers a valuable opportunity for developers, data scientists, and domain experts working on recommender systems to gain a deeper understanding of the system's behavior, identify potential errors or shortcomings, and suggest improvements. This section elaborates on how explainable AI facilitates debugging and model improvement in recommender systems, enhancing their performance and ensuring that users receive relevant and accurate recommendations [28].

2.4 Transparency for System Understanding

One of the primary benefits of explainable AI is its ability to make the decision-making processes of recommender systems more transparent. With a clear understanding of how the system processes input data, weighs various features, and arrives at recommendations, developers can gain insights into the system's strengths and weaknesses. This transparency is crucial for identifying areas of improvement, such as mitigating biases, addressing over- or underfitting, or refining feature engineering.

2.5 Identifying and Addressing Biases

Biases in recommender systems can arise from various sources, such as biased input data, inadequate feature representation, or flawed algorithms. These biases can lead to unfair or unrepresentative recommendations, potentially alienating certain user groups and negatively impacting the overall user experience. Explainable AI can help developers identify biases in the recommender system by providing insights into the factors that contribute to recommendations. By understanding the system's decision-making processes, developers can detect potential biases and implement corrective measures to ensure fairness and diversity in recommendations.

2.6 Error Detection and Correction

Recommendation errors can occur due to various reasons, such as algorithmic flaws, insufficient data, or inadequate feature representation. Explainable AI can help developers pinpoint the sources of these errors by providing detailed explanations of the system's decision-making processes. By understanding how the system processes data and makes recommendations, developers can identify and correct errors, improving the system's overall accuracy and effectiveness.

2.7 Feature Engineering and Selection

Feature engineering and selection play a critical role in the performance of recommender systems. Explainable AI can provide insights into the importance and impact of different features on recommendations, helping developers identify redundant or irrelevant features and focus on the most informative ones. By refining feature engineering and selection, developers can create more efficient and accurate recommender systems that better cater to users' needs and preferences.

2.8 Iterative Model Improvement

Explainable AI supports an iterative model improvement process, allowing developers to continuously refine and optimize recommender systems. By analyzing the explanations provided by the system, developers can identify potential areas of improvement and test different approaches to address them. This feedback loop is essential for ensuring that the recommender system remains effective and relevant as user preferences, market trends, and product offerings evolve over time.

2.9 Collaborative Development and Communication

Explainable AI facilitates collaboration between developers, domain experts, and other stakeholders by offering a common language for discussing and understanding the recommender system's behavior. Clear explanations of the system's decision-making processes enable more effective communication and problem-solving, as stakeholders can identify issues, suggest improvements, and evaluate the impact of changes on the system's performance. This collaborative development approach is crucial for creating more robust, efficient, and user-centric recommender systems. Ultimately, the integration of explainable AI into recommender systems greatly enhances the debugging and model improvement process. By providing transparent insights into the system's behavior, developers can identify and address biases, detect and correct errors, refine feature engineering and selection, and engage in iterative model improvement. Furthermore, explainable AI fosters collaborative development and communication among stakeholders, enabling the creation of more effective and user-centric recommender systems that deliver relevant and accurate recommendations.

3 Types of Explanations in Recommender Systems

To enhance the effectiveness and transparency of recommender systems, explainable AI can provide various types of explanations that cater to different aspects of the decision-making process [29, 30]. By understanding the unique characteristics of each explanation type, businesses can design and implement recommender systems that offer more comprehensive and meaningful insights to users behaviors [31, 32]. In this section, we will explore content-based explanations, collaborative-based explanations, and hybrid explanations in greater detail.

3.1 Content-Based Explanations

Content-based explanations are generated based on the inherent features of the recommended items [33, 34]. These explanations offer insights into the rationale behind recommendations by highlighting the similarities between the suggested items and the user's preferences or past interactions [35, 36]. For example, a content-based explanation may emphasize that a recommended product shares key characteristics with items the user has previously purchased or expressed interest in.

Content-based explanations can be particularly effective for:

- *New Users*: When there is limited data available about a user's preferences, content-based explanations can provide relevant recommendations based on the user's initial interactions with the platform.
- *Diverse Interests*: For users with a wide range of interests, content-based explanations can help identify and recommend items that cater to their specific preferences, enhancing the personalization of the recommender system.
- *Explainability:* Content-based explanations can provide more transparent insights into the decision-making process, as they directly relate to the features of the recommended items, making it easier for users to understand the rationale behind the suggestions.

3.2 Collaborative-Based Explanations

Collaborative-based explanations rely on the preferences and behavior of other users with similar tastes to generate recommendations [37]. These explanations emphasize the social aspect of decision-making by drawing connections between the recommended items and the preferences of like-minded users [38]. A typical collaborative-based explanation might state, "Users who bought this item also bought these items," or "Users with similar preferences enjoyed this product."

Collaborative-based explanations offer several benefits, including:

- *Serendipity*: By leveraging the preferences of other users, collaborative-based explanations can introduce users to new items they may not have discovered otherwise, fostering a sense of serendipity and delight.
- *Social Validation*: These explanations provide a sense of social validation for the recommendations, as users can see that others with similar tastes have found the suggested items valuable or enjoyable.
- *Scalability:* Collaborative-based explanations can scale effectively as the user base grows, making them suitable for large-scale e-commerce platforms.

3.3 Hybrid Explanations

Hybrid explanations combine the strengths of both content-based and collaborative-based explanations to offer a more comprehensive understanding of the recommendation's rationale. These explanations integrate insights from the item's features and the preferences of similar users, providing users with a richer context for the recommendations they receive [39–41].

The advantages of hybrid explanations include:

- *Enhanced Personalization*: By incorporating both content-based and collaborative-based insights, hybrid explanations can cater to a broader range of user preferences and interests, resulting in a more personalized and relevant user experience.
- *Robustness:* Hybrid explanations can be more robust and accurate, as they consider multiple factors in the decision-making process, reducing the likelihood of generating recommendations based on spurious correlations or biases.
- *Flexibility:* The combination of content-based and collaborative-based explanations allows for flexibility in addressing the unique challenges and requirements of different e-commerce platforms and user segments.

In summary, content-based, collaborative-based, and hybrid explanations each offer unique insights into the decision-making processes of recommender systems. By understanding the nuances of these explanation types, businesses can design more transparent and effective recommender systems that cater to the diverse needs and preferences of their users.

4 Techniques for Generating Explanations

4.1 Model-Agnostic Methods

Model-agnostic methods for generating explanations in recommender systems are independent of the underlying model and focus on understanding the input–output relationship. These techniques can be applied to various types of recommendation models, including collaborative filtering, content-based filtering, and hybrid approaches [42]. This versatility makes model-agnostic methods particularly appealing when working with complex, diverse, or proprietary models where the internal workings may be difficult to interpret or access. In this section, we will discuss two widely used model-agnostic methods, namely Local Interpretable Model-agnostic Explanations (LIME) and Shapley Additive Explanations (SHAP).

i. **Local Interpretable Model-agnostic Explanations (LIME)**

LIME is a popular model-agnostic technique designed to provide local explanations for individual predictions made by a black-box model [43]. The fundamental idea behind LIME is to approximate the complex model with a simpler, interpretable model in the vicinity of a specific data point (i.e., a recommendation in the context of recommender systems) [44].

To generate an explanation using LIME, the following steps are performed:

- *Perturbation:* Create a dataset of perturbed instances around the data point of interest (the recommendation to be explained). This can be achieved by sampling from the feature space or using other data manipulation techniques.
- *Model Prediction:* Obtain predictions from the black-box model for each perturbed instance.
- *Weight Assignment:* Compute the similarity between the original data point and each perturbed instance. This similarity is used as a weight, emphasizing the importance of instances that are closer to the data point in question.
- *Model Fitting:* Train a simple, interpretable model (e.g., linear regression) on the perturbed dataset, using the weights from the previous step. This local model serves as an approximation of the black-box model in the vicinity of the data point.
- *Interpretation:* Extract feature importance from the interpretable model to generate an explanation for the recommendation.

By providing local explanations, LIME allows users to understand the factors driving a specific recommendation, even when the overall model is highly complex and non-linear [45].

ii. **Shapley Additive Explanations (SHAP)**

SHAP is another model-agnostic method that provides explanations in the form of feature importances. It is based on the concept of Shapley values, originating from cooperative game theory [46]. Shapley values offer a fair distribution of the payoff (in this case, the prediction) among the features (players) based on their contributions to the overall output.

To compute SHAP values for a recommendation, the following steps are performed:

- *Enumerate Coalitions:* Enumerate all possible feature coalitions (subsets of features). Each coalition represents a different combination of features contributing to the prediction.
- *Model Prediction:* Calculate the prediction for each coalition using the black-box model.
- *Shapley Value Calculation:* Compute the Shapley values for each feature by averaging the marginal contributions of the feature across all possible coalitions.

By calculating the SHAP values, we obtain a measure of each feature's importance in generating a specific recommendation. These values can then be used to create an explanation that offers insights into the factors contributing to the recommendation, fostering transparency and understanding [47]. To sum up, model-agnostic methods like LIME and SHAP provide valuable insights into the recommendations generated by black-box models, enhancing the explainability and interpretability of recommender systems. By offering local, instance-specific explanations, these techniques allow users to understand the rationale behind individual recommendations and facilitate trust and confidence in the recommender system.

4.2 Model-Specific Methods

Model-specific methods for generating explanations in recommender systems are tailored to the underlying recommendation models, taking advantage of their unique properties and structures to provide fine-grained insights into the decision-making processes [48]. By leveraging the intricacies of specific models, these techniques can offer more detailed and accurate explanations for recommendations [14–16, 18, 22]. In this section, we will explore various model-specific methods and their application in explainable recommender systems.

- *Matrix Factorization:* Matrix factorization is a widely used technique for collaborative filtering in recommender systems. It decomposes a user-item interaction matrix into two lower-dimensional matrices representing latent factors for users and items. These latent factors capture the underlying patterns and relationships between users and items, allowing the recommender system to make predictions based on these factors [20, 21]. Explainable matrix factorization techniques can generate explanations by identifying the most important latent factors contributing to a recommendation. For example, if a user has a strong preference for a specific latent factor (e.g., a particular movie genre), and an item exhibits high values for that factor, it can be inferred that the recommendation is based on the user's affinity for that genre. By exposing these latent factors to users, the recommender system can provide a clear rationale for the recommendations, increasing transparency and trust [19–26].

- *Neural Networks:* Deep learning-based recommender systems, such as those built using neural networks, have gained popularity due to their ability to model complex and non-linear relationships between users and items [1–4, 10]. However, the inherent complexity of neural networks can make them challenging to interpret. To generate explanations for neural network-based recommender systems, researchers have developed techniques such as attention mechanisms and layer-wise relevance propagation. Attention mechanisms allow the model to weigh the importance of different input features when making recommendations, effectively highlighting the key factors driving a particular decision. Layer-wise relevance propagation, on the other hand, traces the flow of information through the network, identifying the contribution of individual neurons and features to the final recommendation. By employing these techniques, developers can offer users insights into the reasoning behind the system's decisions, enhancing transparency and interpretability.
- *Knowledge Graphs:* Knowledge graph-based recommender systems utilize structured knowledge about items, users, and their relationships to generate recommendations. These systems can exploit the rich information encoded in knowledge graphs to provide detailed and context-aware explanations. To generate explanations in knowledge graph-based recommender systems, techniques such as pathfinding and subgraph extraction can be employed. Pathfinding algorithms identify the most relevant paths connecting a user and a recommended item in the knowledge graph, revealing the relationships that contribute to the recommendation [41–45]. Subgraph extraction techniques, on the other hand, extract relevant portions of the knowledge graph that provide context for the recommendation, offering users a comprehensive understanding of the rationale behind the suggestion.

To sum up, model-specific methods for generating explanations in recommender systems harness the unique properties of various recommendation models to provide detailed and accurate insights into the decision-making processes. By employing techniques such as explainable matrix factorization [20, 21], attention mechanisms, and knowledge graph-based approaches, developers can create transparent and interpretable recommender systems that foster trust, user confidence, and engagement. As AI models continue to evolve and become more sophisticated, the importance of model-specific explainability methods in the design and implementation of recommender systems will only grow.

4.3 Practical Implementations of Explainable Recommender Systems

4.3.1 Case Study: Amazon

Amazon, one of the largest and most successful e-commerce platforms globally, has long recognized the value of personalized recommendations in driving customer satisfaction and increasing revenue. To ensure that users can understand and trust the rationale behind the recommendations they receive, Amazon has integrated explainable AI into its recommender system [20–25].

One notable feature of Amazon's explainable recommender system is the "Why Recommended" option. When users hover over or click on this feature, they are provided with insights into the factors that contributed to the recommendation. These insights can take various forms, such as:

- *Content-based explanations:* Amazon might explain that a recommended product is similar to a previously purchased item or is related to items in the user's browsing history. For example, if a user has recently purchased a smartphone, Amazon might recommend a phone case or screen protector and explain that the recommendation is based on the user's recent purchase.
- *Collaborative-based explanations:* Amazon may also provide explanations based on the behavior of other users with similar tastes or interests. For instance, the platform might recommend a book to a user and explain that the suggestion is based on the purchasing history of other users who have bought similar books.
- *Personalized explanations:* In some cases, Amazon may offer personalized explanations that take into account the user's unique preferences and interests. These explanations can be particularly effective in building trust, as they demonstrate that the recommender system understands and caters to the user's specific needs.
- *Context-aware explanations:* Amazon's explainable recommender system also considers the context in which users interact with the platform. For example, the system may generate different explanations for recommendations made during holiday seasons or promotional events, reflecting the unique circumstances that may influence users' preferences and needs at these times.

By providing these various explanations, Amazon's recommender system effectively fosters trust and user confidence in the recommendations (Fig. 1). Customers can understand the rationale behind the suggestions they receive and feel more assured that the recommended items are relevant and tailored to their preferences. Moreover, the integration of explainable AI into Amazon's recommender system offers additional benefits beyond enhancing trust and user confidence. For instance, the insights provided by the explainable AI can help the platform's developers and domain experts identify potential biases or shortcomings in the system, allowing them to refine and optimize the recommendation algorithms. Additionally, the system's transparency can help Amazon demonstrate compliance with ethical guidelines and

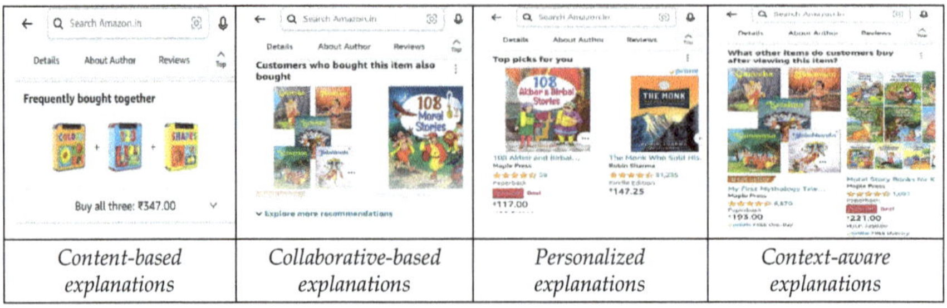

Fig. 1 Amazon's explainable recommender system displaying the "why recommended" features on the user interface (*Source* Author)

legal regulations that emphasize the importance of fairness and accountability in AI systems.

Therefore, Amazon's implementation of explainable AI in its recommender system serves as an excellent example of how e-commerce platforms can leverage transparency and interpretability to enhance user trust and confidence in their recommendations. By providing users with clear, understandable explanations for the suggestions they receive, Amazon can ensure that its customers feel confident in their choices and are more likely to engage with the recommended products or services. As a result, Amazon's explainable recommender system not only improves user satisfaction but also contributes to the company's continued success and growth in the highly competitive e-commerce market.

4.3.2 Case Study: Netflix

Netflix, a leading streaming service provider, has successfully integrated explainable AI into its recommender system to enhance user engagement and satisfaction. With millions of subscribers and a vast library of movies, TV shows, and documentaries, providing personalized and relevant recommendations is crucial for retaining users and encouraging them to explore new content [35]. This section examines how Netflix leverages explainable AI to provide transparent and meaningful recommendations that resonate with its users. To understand the role of explainable AI in Netflix's recommender system, it is essential to first comprehend the underlying mechanics of the system. The system employs a combination of collaborative filtering, content-based filtering, and deep learning techniques to analyze user behavior, preferences, and content metadata [1, 2, 14, 15]. By combining these approaches, Netflix can generate personalized recommendations that cater to each user's unique tastes and viewing habits. Explainable AI comes into play by providing users with clear explanations for the recommendations they receive. Netflix displays these explanations directly on the user interface, offering insights into the rationale behind each suggestion. Some common explanations include:

- *"Because you watched [movie/series]":* This explanation highlights the connection between a user's viewing history and the recommended content. By showcasing the similarities between the user's past choices and the suggested item, Netflix demonstrates the relevance of the recommendation and encourages users to explore the content further.
- *"Top picks for [user]":* These recommendations are specifically curated for each user based on their viewing history, preferences, and behavior on the platform. By explicitly stating that the recommendations are tailored for the user, Netflix builds trust and confidence in the system's ability to understand and cater to individual tastes.
- *"Trending Now":* This category highlights popular content among the broader Netflix user base. While not as personalized as other recommendations, it still provides context for the suggestion, allowing users to understand the rationale behind the inclusion of the content in their recommendations.
- *"Because you added [movie/series] to My List":* Netflix takes into account users' actions, such as adding content to their watchlist, to generate recommendations. By explaining that the suggestions are based on the user's expressed interest in similar content, Netflix reinforces the relevance of the recommendations.
- *"Critically acclaimed [genre] films":* This explanation leverages external sources, such as critical reviews or awards, to provide context for the recommendation. By associating the recommended content with positive feedback from critics or recognition within the industry, Netflix bolsters the perceived quality and relevance of the suggestion (Fig. 2).

By integrating explainable AI into its recommender system, Netflix effectively enhances user trust, engagement, and satisfaction. Users can better understand the rationale behind each recommendation, which fosters a sense of trust and confidence in the system's ability to curate content that aligns with their preferences [23]. Moreover, explainable AI helps users discover new content they might not have found otherwise, thereby encouraging exploration and increasing the overall value of the platform [15, 16, 50] and [51]. In conclusion, Netflix serves as a prime example of how explainable AI can be successfully integrated into recommender systems to provide transparent and meaningful recommendations that resonate with users. By

"Top picks for [user]"	*"Critically acclaimed TV shows"*	*"Trending Now"*	*"Because you watched [movie/series]"*	*"Because you added [movie/series] to My List*

Fig. 2 Netflix displays explanations directly on the user interface. (*Source* Author)

offering clear explanations for each suggestion, Netflix fosters trust and confidence in its recommender system, ultimately leading to increased user engagement and satisfaction.

5 Conclusion and Future Recommendation

In an increasingly competitive e-commerce landscape, the adoption of explainable AI in recommender systems has become paramount for businesses seeking to differentiate themselves and provide the best possible user experience. The integration of explainable AI not only fosters trustworthiness and user confidence but also ensures that recommender systems operate in an ethical and legally compliant manner, while facilitating debugging and continuous improvement of these systems. In this concluding section, we will revisit the key benefits of explainable AI in recommender systems and explore the implications and prospects of this technology.

Firstly, enhancing trustworthiness and user confidence is a crucial advantage of incorporating explainable AI into recommender systems. When users understand the rationale behind recommendations, they are more inclined to trust the system and engage with the suggested products or services. This increased trust results in higher conversion rates and improved customer satisfaction, which are essential for e-commerce platforms' success. By providing users with personalized, dynamic, and consistent explanations, explainable AI enables them to make more informed decisions and feel confident in their choices, ultimately leading to stronger customer relationships and increased loyalty.

Secondly, explainable AI in recommender systems helps ensure ethical and legal compliance. As AI systems gain greater autonomy and influence, concerns about fairness, accountability, and transparency have become more pressing. Explainable AI can identify and mitigate biases and unfairness within the system by offering insights into the factors contributing to a recommendation, ensuring that the recommender system adheres to ethical principles. From a legal standpoint, incorporating explainable AI into recommender systems enables businesses to comply with regulations such as GDPR, which emphasize transparency and accountability in AI systems.

Thirdly, explainable AI plays a vital role in debugging and improving recommender systems. By making the decision-making processes more transparent, developers, domain experts, and other stakeholders can identify potential errors, biases, or areas of improvement in the system. This feedback loop allows for the optimization of recommender systems, ensuring that they continue to provide relevant and accurate recommendations to users.

Looking towards the future, the role of explainable AI in recommender systems is expected to grow as AI systems become more complex and users demand greater transparency. The continued development and refinement of explainable AI techniques will lead to more sophisticated and accurate explanations, further enhancing user trust and confidence in recommender systems. In addition, as ethical and legal concerns surrounding AI become more prominent, the integration of explainable AI will become a crucial component in maintaining compliance and demonstrating a commitment to responsible AI practices. Furthermore, as the field of explainable AI advances, we can expect the emergence of new techniques and approaches that will enable even greater transparency and interpretability in recommender systems. This ongoing innovation will likely lead to more effective and engaging recommendation experiences for users, benefiting both consumers and e-commerce businesses alike. In conclusion, explainable AI is an indispensable component of recommender systems in the e-commerce domain. It plays a critical role in fostering trustworthiness, ensuring ethical and legal compliance, and facilitating the debugging and improvement of these systems. As AI continues to evolve and become an integral part of our daily lives, explainability will remain a crucial aspect of designing and implementing recommender systems that effectively meet users' needs and expectations.

References

1. Abdollahi, B., & Nasraoui, O. (2016). Explainable matrix factorization for collaborative filtering. In Proceedings of the 25th International Conference Companion on World Wide Web (pp. 5–6). https://doi.org/10.1145/2872518.2889402.
2. Abdollahi, B., & Nasraoui, O. (2018). Transparent recommendations: An approach to explainable recommender systems. In Proceedings of the 12th ACM Conference on Recommender Systems (pp. 364–365). https://doi.org/10.1145/3240323.3240375.
3. Adadi, A., & Berrada, M. (2018). Peeking inside the black-box: A survey on explainable artificial intelligence (XAI). *IEEE Access, 6*, 52138–52160. https://doi.org/10.1109/ACCESS.2018.2870052
4. Arrieta, A. B., Díaz-Rodríguez, N., Del Ser, J., Bennetot, A., Tabik, S., Barbado, A., Herrera, F., et al. (2020). Explainable Artificial Intelligence (XAI): Concepts, taxonomies, opportunities and challenges toward responsible AI. *Information Fusion, 58*, 82–115. https://doi.org/10.1016/j.inffus.2019.12.012
5. Bhatt, U., Xiang, A., Sharma, S., Weller, A., Taly, A., Jia, Y., & Eckersley, P., et al. (2020). Explainable machine learning in deployment. In Proceedings of the 2020 Conference on Fairness, Accountability, and Transparency (pp. 648–657). https://doi.org/10.1145/3351095.3372850.
6. Bilgic, M., & Mooney, R. J. (2005). Explaining recommendations: Satisfaction vs. promotion. In Proceedings of Beyond Personalization 2005: A Workshop on the Next Stage of Recommender Systems Research at the 2005 International Conference on Intelligent User Interfaces (pp. 13–18).
7. Biren, O., & Cotton, C. (2017). Explanation and justification in machine learning: A survey. In Proceedings of the IJCAI-17 Workshop on Explainable Artificial Intelligence (XAI) (pp. 1–9).
8. Bobadilla, J., Ortega, F., Hernando, A., & Gutiérrez, A. (2013). Recommender systems survey. *Knowledge-Based Systems, 46*, 109–132. https://doi.org/10.1016/j.knosys.2013.03.012
9. Chen, T., & Guestrin, C. (2016). XGBoost: A scalable tree boosting system. In Proceedings of the 22nd ACM SIGKDD International Conference on Knowledge Discovery and Data Mining (pp. 785–794). https://doi.org/10.1145/2939672.2939785.
10. Gaur, L., & Sahoo, B. M. (2022). Introduction to explainable AI and intelligent transportation. *Explainable artificial intelligence for intelligent transportation systems: Ethics and applications* (pp. 1–25). Springer International Publishing.
11. Gilpin, L. H., Bau, D., Yuan, B. Z., Bajwa, A., Specter, M., & Kagal, L. (2018). Explaining explanations: An approach to evaluating interpretability of machine learning. arXiv preprint arXiv:1806.00069.

12. Gonzalez, M. F., Liu, W., Shirase, L., Tomczak, D. L., Lobbe, C. E., Justenhoven, R., & Martin, N. R. (2022). Allying with AI? Reactions toward human-based, AI/ML-based, and augmented hiring processes. *Computers in Human Behavior, 130*, 107179.
13. Gramegna, A., & Giudici, P. (2021). SHAP and LIME: An evaluation of discriminative power in credit risk. *Frontiers in Artificial Intelligence, 4*, 752558.
14. Guidotti, R., Monreale, A., Pedreschi, D., & Giannotti, F. (2021). Principles of explainable artificial intelligence. *Explainable AI Within the Digital Transformation and Cyber Physical Systems: XAI Methods and Applications*, 9–31.
15. Gupta, V., & Sahu, G. (2021). Reviving the Indian hospitality industry after the Covid-19 pandemic: The role of innovation in training. *Worldwide hospitality and tourism themes, 13*(5), 599–609.
16. Gupta, V., Roy, H., & Sahu, G. (2022). HOW the tourism & hospitality lecturers coped with the transition to online teaching due to COVID-19: An assessment of stressors, negative sentiments and coping strategies. *Journal of Hospitality, Leisure, Sport and Tourism Education, 30*, 100341.
17. Haque, A. B., Islam, A. N., & Mikalef, P. (2023). Explainable Artificial Intelligence (XAI) from a user perspective: A synthesis of prior literature and problematizing avenues for future research. *Technological Forecasting and Social Change, 186*, 122120.
18. Jannach, D., & Adomavicius, G. (2016). Recommendations with a purpose. In Proceedings of the 10th ACM Conference on Recommender Systems (pp. 7–10). https://doi.org/10.1145/2959100.2959176.
19. Jiang, J., Kahai, S., & Yang, M. (2022). Who needs explanation and when? Juggling explainable AI and user epistemic uncertainty. *International Journal of Human-Computer Studies, 165*, 102839.
20. Jiarpakdee, J., Tantithamthavorn, C. K., Dam, H. K., & Grundy, J. (2020). An empirical study of model-agnostic techniques for defect prediction models. *IEEE Transactions on Software Engineering, 48*(1), 166–185.
21. Koren, Y., Bell, R., & Volinsky, C. (2009). Matrix factorization techniques for recommender systems. *Computer, 42*(8), 30–37. https://doi.org/10.1109/MC.2009.263
22. Langer, M., Oster, D., Speith, T., Hermanns, H., Kästner, L., Schmidt, E., Sesing, A., & Baum, K. (2021). What do we want from Explainable Artificial Intelligence (XAI)?–A stakeholder perspective on XAI and a conceptual model guiding interdisciplinary XAI research. *Artificial Intelligence, 296*, 103473.
23. Lee, S. (2022). AI as an explanation agent and user-centered explanation interfaces for trust in AI-based systems. In *Human-Centered Artificial Intelligence* (pp. 91–102). Academic Press.
24. Linden, G., Smith, B., & York, J. (2003). Amazon.com recommendations: Item-to-item collaborative filtering. IEEE Internet Computing, 7(1), 76–80. https://doi.org/10.1109/MIC.2003.1167344
25. Lundberg, S. M., & Lee, S. I. (2017). A unified approach to interpreting model predictions. In Proceedings of the 31st International Conference on Neural Information Processing Systems (pp. 4768–4777).
26. Markus, A. F., Kors, J. A., & Rijnbeek, P. R. (2021). The role of explainability in creating trustworthy artificial intelligence for health care: A comprehensive survey of the terminology, design choices, and evaluation strategies. *Journal of Biomedical Informatics, 113*, 103655.
27. Meske, C., Abedin, B., Klier, M., & Rabhi, F. (2022). Explainable and responsible artificial intelligence. *Electronic Markets*, 1–4.
28. Miller, T., Howe, P., & Sonenberg, L. (2017). Explainable AI: Beware of inmates running the asylum or: How I learnt to stop worrying and love the social and behavioural sciences. arXiv preprint arXiv:1712.00547.
29. Montavon, G., Samek, W., & Müller, K. R. (2018). Methods for interpreting and understanding deep neural networks. *Digital Signal Processing, 73*, 1–15. https://doi.org/10.1016/j.dsp.2017.10.011
30. Nunes, I., & Jannach, D. (2017). A systematic review and taxonomy of explanations in decision support and recommender systems. *User Modeling and User-Adapted Interaction, 27*(3–5), 393–444. https://doi.org/10.1007/s11257-017-9195-0

31. Paz-Ruza, J., Eiras-Franco, C., Guijarro-Berdiñas, B., & Alonso-Betanzos, A. (2022). Sustainable personalisation and explainability in dyadic data systems. *Procedia Computer Science, 207*, 1017–1026.
32. Quadrianto, N., Schuller, B. W., & Lattimore, F. R. (2021). Ethical machine learning and artificial intelligence. *Frontiers in Big Data, 4*, 742589.
33. Ribeiro, M. T., Singh, S., & Guestrin, C. (2016). "Why should I trust you?": Explaining the predictions of any classifier. In Proceedings of the 22nd ACM SIGKDD International Conference on Knowledge Discovery and Data Mining (pp. 1135–1144). https://doi.org/10.1145/2939672.2939778
34. Saeed, W., & Omlin, C. (2023). Explainable ai (xai): A systematic meta-survey of current challenges and future opportunities. *Knowledge-Based Systems*, 110273.
35. Sahu, G., Gaur, L., & Singh, G. (2021). Applying niche and gratification theory approach to examine the users' indulgence towards over-the-top platforms and conventional TV. *Telematics and Informatics, 65*, 101713.
36. Sahu, G., Gaur, L., & Singh, G. (2022, November). Analyzing the Users' De-familiarity with Thumbnails on OTT Platforms to Influence Content Streaming. In *2022 International Conference on Computing, Communication, and Intelligent Systems (ICCCIS)* (pp. 551–556). IEEE.
37. Vale, D., El-Sharif, A., & Ali, M. (2022). Explainable artificial intelligence (XAI) post-hoc explainability methods: Risks and limitations in non-discrimination law. *AI and Ethics*, 1–12.
38. Weber, P., Carl, K. V., & Hinz, O. (2023). Applications of explainable artificial intelligence in finance—A systematic review of finance, information systems, and computer science literature. *Management Review Quarterly*, 1–41.
39. Yalcin, O. G. (2021). GDPR compliant data processing and privacy preserving technologies: A literature review on notable Horizon 2020 projects. *New Trends in Disruptive Technologies, Tech Ethics and Artificial Intelligence: The DITTET Collection*, 166–177.
40. Zhang, S., Yao, L., Sun, A., & Tay, Y. (2019). Deep learning-based recommender system: A survey and new perspectives. *ACM Computing Surveys (CSUR), 52*(1), 1–38. https://doi.org/10.1145/3285029
41. Zhang, Y., & Chen, X. (2020). Explainable recommendation: A survey and new perspectives. *Foundations and Trends® in Information Retrieval, 14*(1), 1–101. https://doi.org/10.1561/1500000066
42. Zimmermann, R., Mora, D., Cirqueira, D., Helfert, M., Bezbradica, M., Werth, D., Weitzl, W. J., Riedl, R., & Auinger, A. (2023). Enhancing brick-and-mortar store shopping experience with an augmented reality shopping assistant application using personalized recommendations and explainable artificial intelligence. *Journal of Research in Interactive Marketing, 17*(2), 273–298. https://doi.org/10.1108/JRIM-09-2021-0237
43. Gaur, L., Sahoo, B. M. (2022). Intelligent transportation technology enablers. In: Explainable Artificial Intelligence for Intelligent Transportation Systems. Springer. https://doi.org/10.1007/978-3-031-09644-0_2.
44. Gaur, L., Ratta, M., & Gaur, A. (2022). Future of DeepFakes and Ectypes In: Deepfakes. CRC Press, 9781003231493.
45. Gaur, L., & Sahoo, B. M. (2022). Explainable AI in ITS: Ethical concerns. In: Explainable Artificial Intelligence for Intelligent Transportation Systems. Springer. https://doi.org/10.1007/978-3-031-09644-0_5.
46. Gaur, L., Sahoo, B.M. (2022). Intelligent transportation system: Modern business models. In: Explainable Artificial Intelligence for Intelligent Transportation Systems. Springer. https://doi.org/10.1007/978-3-031-09644-0_4.
47. Gaur, L., Jhanjhi, N. Z., Bakshi, S., & Gupta, P. (2022). Analyzing consequences of artificial intelligence on jobs using topic modeling and keyword extraction. In *2022 2nd International Conference on Innovative Practices in Technology and Management (ICIPTM)* (pp. 435–440). https://doi.org/10.1109/ICIPTM54933.2022.9754064.
48. Gaur, L., Bhandari, M., Razdan, T., Mallik, S., & Zhao, Z. (2022). Explanation-driven deep learning model for prediction of brain Tumour status using MRI image data. *Frontier in Genetics, 13*, 822666.

49. Anshu, K., Gaur, L., & Singh, G. (2021). Co-creation: Interface for online affective experience and repurchase intention. *International Journal of Business and Economics*, *20*(2), 161–185. ISSN 1607–0704.
50. Sharma, S., Singh, G., Gaur, L., & Afaq, A. (2022). Exploring customer adoption of autonomous shopping systems. *Telematics and Informatics*, *73*, 101861, ISSN 0736–5853. https://doi.org/10.1016/j.tele.2022.101861.
51. Sahu, G., Singh, G., Singh, G. & Gaur, L. (2024). Exploring new dimensions in OTT consumption: an empirical study on perceived risks, descriptive norms and goal-directed behaviour. *Asia Pacific Journal of Marketing and Logistics*. https://doi.org/10.1108/APJML-07-2023-0690.

"The AI Revolution in E-Commerce: Personalization and Predictive Analytics"

Chitra Krishnan and Jasmine Mariappan

Abstract E-commerce, has totally changed the way we purchase goods and do business. Customers may now browse, purchase, and receive goods all from the comfort of their own homes, thanks to advancements in technology and the expansion of the internet. Over the years, e-commerce has undergone significant transformation, affecting both the retail environment and customer expectations. Integration of artificial intelligence (AI) into the e-commerce sector, especially in the customization domain, is one important component that has made a substantial contribution to this progress. The volume of data produced by online shoppers increased quickly alongside the growth of e-commerce. Traditional techniques of customization could no longer meet the wide range of requests and tastes of consumers. E-commerce customization started to heavily rely on AI algorithms, especially machine learning. Through extensive data analysis, artificial intelligence (AI) has the potential to discern individual user habits, preferences, and behaviors, therefore empowering e-commerce platforms to provide highly customized purchasing experiences. A more customized and relevant buying experience for every client resulted from this, replacing the previous one-size-fits-all strategy. This chapter demonstrates how big data may be used by e-commerce companies to provide highly customized products that meet the demands of their customers and provide them a competitive advantage.

Keywords Artificial intelligence · e-commerce · Consumer behavior · Personalization · Predictive analytics

---

C. Krishnan (✉)
Symbiosis Centre for Management Studies, Symbiosis International Deemed University, Noida, India
e-mail: ananya.chitra@gmail.com

J. Mariappan
University of Technology and Applied Sciences, Ibra, Oman

L. Gaur and A. Abraham (eds.), *Role of Explainable Artificial Intelligence in E-Commerce*, Studies in Computational Intelligence 1094,
https://doi.org/10.1007/978-3-031-55615-9_4

1 Introduction

By predicting buying patterns based on the products that customers buy and when they get them, artificial intelligence in online shopping is revolutionizing the e-commerce sector. In the event, for example, that weekly online shoppers buy a particular brand of rice, the online retailer may then send these shoppers a tailored offer for that product or even use artificial intelligence (AI) to suggest a useful item that pairs well with rice dishes. AI-enabled internet-based business tools, such as the Google Duplex tool, are developing capabilities such as establishing standard records (using the user's natural voice) and, in any case, submitting online purchase orders on their behalf. Consumers' purchasing decisions and behavioural habits are significantly impacted by a variety of other factors in the period of Industrial Revolution 4.0 in addition to the tremendous advancements in technology. One of the cornerstones of strength for any successful retailer is customer satisfaction. As a result, the introduction of AI makes it easier to comprehend and forecast consumer behavior patterns and trends in e-commerce. This is because using analytics alone was unable to accurately demonstrate the behavioural patterns because of the enormous volume and variance of data, which opened the door for AI's entry into the picture. When combined with analytics, AI can unleash its value and simplify.

One of the most important tools in the invention of e-commerce is data. The benefits to the entire society may be summed up as follows: merchants are able to better understand the demands of their customers; consumers are able to get the exact product they want; and the government is able to evaluate the economic effect of e-commerce. Since customers cannot try on or handle things in traditional locations, the digital revolution over the last five years has demonstrated the necessity to offer more unique services. Over the past several years, AI has become more and more prevalent, impacting a wide range of enterprises, but e-commerce has been particularly hard hit.

AI refers to computer programs that simulate human cognitive processes. It involves creating robots that are capable of cognitive tasks that were previously limited to human performance, such speech recognition, visual perception, problem-solving, and decision-making. Computer technology have significantly enhanced the interactions between businesses and their customers.

Generative AI is being used by e-commerce companies to scale up the generation of their marketing assets and customize it for various consumers. A copywriter may, for instance, use generative AI to tailor a marketing email to different consumer categories after it has been written. Additionally, marketers may use generative AI to provide input on their positioning and brand message to make sure it fits with the personas of their target audience.

AI is the term used to describe computers that respond to inputs in a way that is consistent with human responses in the past, given human intellect, judgment, and intention. The current economy depends on AI systems that can effectively meet customer expectations across a variety of businesses [1]. The goal of AI is to develop methods for systems that display aspects of intelligent behavior. These technologies are meant to simulate human sensory and mental processes [2].

E-commerce, sometimes referred to as electronic commerce, is the term used to describe the services and activities related to online buying and selling of products and services [3, 4]. Businesses are using e-commerce increasingly often as a result of the rising demand for online services and their capacity to provide them a competitive edge [5–7]. The foundation of e-commerce is a strategy based on electronic hardware and organizational innovation, which devices the business cycles of all business exercises. These include not only the remotely arranged business measures in business exercises of endeavours, like network showcasing, electronic installment, coordination, and circulation, but also the internal business cycles of endeavors, like supply chain management (SCM), enterprise resource planning (ERP), management information system (MIS), customer relationship management (CRM), and human resource management (HRM).

Even though e-commerce is rapidly developing and becoming more affordable, businesses still have difficulties with it. Companies need to continuously adjust their business strategies to meet the evolving requirements of their customers [6]. Depending on the situation, AI can be implemented as a method, system, algorithm, or tool [8]. By using big data to deliver personalized services that precisely meet customers' expectations, artificial intelligence (AI) gives businesses the opportunity to gain a competitive edge in the e-commerce industry [9, 10].

In order to make online sales and purchases of goods and services simpler, AI refers to the application of strategies, tools, frameworks, or algorithms to e-commerce. Approximately 4,000 study publications at the organizational and consumer levels have been published on a range of issues during the previous thirty years of research [11].

Online firms can accomplish higher results with the use of artificial intelligence. As machines grow in intelligence, they can perform physical labor and help people. We can concentrate on more intricate facets of our firm because of how wonderfully these machines are managing their tasks. Artificial Intelligence has patterns, but there are so many more possibilities that it is hard to cover them all in one essay. Let's concentrate on a few of the most stunning and current styles. There are two career paths where traditional human intelligence may not be as useful: supply chain system management and demand prediction [12].

AI is evolving to imitate human cognition in order to meet these new issues in the e-commerce industry [13]. AI assists e-commerce platforms in monitoring and managing its clientele by collecting and analysing diverse data to ensure the delivery of superior services [14].

The purpose of this chapter is to explain how AI is changing the e-commerce business. Chapter goals includes:

- Significance of personalization and predictive analytics in influencing consumer demand.
- Role of AI in optimizing pricing and inventory management.
- Potential of AI-powered supply chain management.

2 Personalization

In the present digital world, personalization has become a crucial part of every successful marketing strategy. AI systems are able to deliver in-the-moment product suggestions by examining a user's browsing history, past purchases, and interactions with the website. The probability of a successful transaction is raised by these recommendations, which are very pertinent and tailored to the customer's tastes.

Personalized content is provided to consumers via e-commerce platforms using artificial intelligence. Personalization increases engagement and increases the likelihood of conversion through landing pages, emails, and banner ads. AI is able to forecast a variety of consumer behavior, including the possibility that a customer will make a purchase, their potential lifetime value, and the items they are most likely to be interested in. E-commerce companies may use this information to maximize their marketing plans and client retention initiatives.

Artificial intelligence (AI)-driven chatbots and virtual assistants provide prompt, customized customer service, answering questions from clients and working around the clock. This lessens the effort for human customer care representatives while also increasing customer happiness [15].

2.1 Benefits of Personalization for e-commerce Businesses and Consumers

Personalization in e-commerce benefits companies as well as consumers. Businesses may gain a lot from personalization, including increased sales, better customer loyalty, and more consumer involvement. Better customer service, more pertinent product recommendations, customized content, improved search results, and more may all be provided to clients through personalization. Few advantages include:

(A) **Benefits for e-commerce businesses:**

- Personalization makes it easier for your e-commerce to target potential customers and to properly nurture them so that they become customers. By using customisation, you may assist your clients in sifting through the noise and locating precisely what they're searching for. An individualised home page, for example, lowers the bounce rate. This is so because the page has been carefully chosen to appeal to the customer's tastes and purchasing patterns, which boosts interaction and conversion rates.
- Personalization features such as product recommendations, customized content, and personalized search results can help to improve customer engagement with an e-commerce platform. Customers are more likely to spend more time on a platform that provides relevant content and recommendations.

- Customers may reduce the number of products they choose by using personalized offers, discounts, and suggestions. A significant portion of consumers—nearly 40%—have abandoned an online retailer due to being overtaken by choice. A website that lets users select material and receive well-chosen suggestions can improve the shopping experience for consumers who purchase a wide range of items from you that appeal to various target audiences and demographics.
- Personalization can help to improve conversion rates by providing customers with a more seamless shopping experience. Customers are more likely to make a purchase if they can easily find the products they are looking for and if they are provided with relevant recommendations.

(B) **Benefits for consumers:**

- Product recommendations and other personalization options can assist customers in locating goods that are most pertinent to their preferences and interests. This can save waiting time and enhance the shopping experience.
- Customized content can provide consumers with a more tailored experience. This can include email campaigns featuring products that are similar to previous purchases, or landing pages that highlight products that are most relevant to the customer.
- Personalized search results can help consumers to find the products they are looking for more quickly and easily. This can save time and reduce frustration.
- Dynamic pricing can benefit consumers by offering them personalized prices based on their behavior and preferences. This can provide consumers with better deals and incentives to make purchases.
- Personalization features such as AI-driven chatbots can help consumers to find what they need and get answers to their questions more quickly and easily. This can improve the overall shopping experience and increase customer satisfaction.

3 Predictive Analytics

Fundamentally, predictive analytics in e-commerce involves employing data, statistical algorithms, and machine learning methods to anticipate future results by analysing past data. Predictive analytics may assist companies in e-commerce and retail by helping them anticipate consumer behavior, optimize their marketing plans, and thereby increase operational effectiveness. Thus, predictive analytics can help Indian e-commerce businesses reduce their logistical costs. By examining past data to predict future demand, businesses may streamline their supply chain and cut down on wasteful spending. Predictive analytics can be applied to e-commerce to comprehend and forecast consumer behavior, preferences, and buying patterns. Predictive analytics is used in e-commerce in the following ways:

- First and foremost, predictive analytics has the potential to greatly improve consumer targeting and comprehension in e-commerce and retail. Companies that analyze customer behavior data can forecast future purchasing trends and adjust their marketing strategies appropriately.
- Pricing strategy may also be optimized with predictive analytics. Companies may optimize profitability by determining the appropriate product pricing by taking into account variables like seasonality, competition, and demand.
- Predictive analytics plays a crucial role in the success of e-commerce businesses by enabling them to understand and predict customer behavior, optimize pricing strategies, and improve overall business operations.
- Businesses can further enhance their inventory management with its assistance. Through forecasting demand, companies may lower the risk of stockouts and overstocking by making sure they have the proper quantity of inventory on hand when needed.
- Additionally, companies may increase lifetime value by anticipating customer attrition and taking proactive measures to keep their clients.
- Predictive analytics may be applied to enhance product suggestions. For example, companies may use consumer behavior and preference data to recommend items that are very likely to be bought by the client.

3.1 Benefits of Predictive Analytics for e-commerce Businesses and Consumers

Predictive analytics is the practise of determining the likelihood of future outcomes based on historical data by employing data, statistical algorithms, and machine learning approaches. Predictive analytics has many advantages for both businesses and customers in the context of e-commerce operations. Among these advantages are:

- For e-commerce companies, predictive analytics has a number of advantages. By using preventative measures based on prediction rates and various applications, it helps businesses to obtain important information for improved decision making [16].
- E-commerce businesses may increase inventory control, create a strong supply chain, identify fraud, customize suggestions, estimate inventory, assess marketing efficacy, and enhance the consumer shopping experience by implementing predictive analytics [17].
- In order to evaluate client loyalty, provide advertising rights, and create marketing plans based on user preferences, it also aids in the analysis of click-flow data [18].
- Predictive analytics is now a critical component of strategic choices that can boost a company's earnings and ensure its long-term viability [19].

4 Integration of Personalization and Predictive Analytics

Personalization and predictive analytics are two powerful technologies that combine to provide customers a more engaging and customized shopping experience. Personalization involves adjusting the purchase process to each customer's specific needs and preferences. To do this effectively, e-commerce businesses need to collect and evaluate data on the preferences and actions of their customers. Information about past purchases, browsing habits, location, and demographics may be included in this. Predictive analytics employs data and statistical algorithms to anticipate future occurrences and patterns of behavior based on past data. Projecting what will be purchased most frequently, gauging the demand for specific goods, and determining which customers will make the most purchases are some examples of this.

By analyzing customer data, e-commerce businesses can use personalization to offer a more pertinent and engaging purchasing experience. Offering personalized promos, creating a unique shopping experience based on a customer's browsing history, and recommending products based on prior purchases are a few examples of how this can be done. Predictive analytics can provide useful information about client behavior to improve personalization. By looking at customer behavior and preference data, e-commerce enterprises can enhance their customization algorithms to give more accurate and applicable recommendations and promotions.

Personalization generates valuable data on customer behavior that can be used to improve predictive analytics. By collecting data on customer preferences, e-commerce businesses can use this information to refine their predictive models, allowing them to make more accurate predictions about which products are most likely to be of interest to their customers.

Together, personalization and predictive analytics can drive sales: By providing a more personalized and engaging shopping experience, e-commerce businesses can increase customer engagement and drive sales. By using predictive analytics to refine their personalization algorithms, e-commerce businesses can provide more accurate and relevant recommendations and promotions, further increasing the likelihood of a sale.

In summary, personalization and predictive analytics complement each other by using data to provide a more personalized and engaging shopping experience for customers. By analyzing customer behavior and preferences data, e-commerce businesses can use personalization to offer more relevant recommendations and promotions. By using predictive analytics to refine their personalization algorithms, e-commerce businesses can provide more accurate and relevant recommendations and promotions, further increasing the likelihood of a sale.

5 Applications of AI in E-commerce

AI has transformed the way e-commerce companies work. AI-powered technologies have been created to assist e-commerce enterprises in automating and streamlining their operations. In order to find out about a product and purchase it online, individuals increasingly rely on voice assistants like Alexa and Siri to know their daily agenda. Here's a rundown of some of the various AI uses in e-commerce:

- Electronic commerce, or e-commerce, has totally changed the way we purchase goods and do business. Customers may now browse, purchase, and receive goods all from the comfort of their own homes, thanks to advancements in technology and the expansion of the internet. Over the years, e-commerce has undergone significant transformation, affecting both the retail environment and customer expectations.
- One significant factor that has contributed significantly to this advancement is the integration of artificial intelligence (AI) into the e-commerce industry, particularly in the customisation space. Along with the fast expansion of e-commerce, so did the amount of data generated by online customers. The vast array of consumer demands and preferences could no longer be satisfied by traditional customisation methods.
- In order to aid your e-commerce firm with customer care, chatbots and virtual assistants may answer questions from customers and make online purchasing easier by offering advice. For understanding and responding to client requests, they employ NLP, AI, and, most recently, generative AI. For company owners, AI has made this process easier. eCommerce companies may now swiftly and automatically create distinctive and dynamic product descriptions thanks to this kind of service. Compared to the traditional method, which can take some time, this is more efficient and takes less time to write a great product description.
- With data analysis, anomaly detection, and real-time transaction monitoring, artificial intelligence (AI) may help with fraud detection and prevention. The system can identify anomalous transactions and flag them for more examination. These transactions can include high-value transfers, several transactions in a brief period of time, or transactions from unknown places.
- AI algorithms analyze customer data such as search history, purchase history, and browsing behavior to offer products that are most likely to be of interest to them. This assist e-commerce enterprise in increasing customer engagement, increasing sales, and decreasing cart abandonment rates. AI-powered systems can also personalize product recommendations based on client preferences and previous behaviors, increasing the likelihood of a sale.
- An artificial intelligence system that creates text, graphics, or other media in response to commands is known as generative AI. DALL-E and ChatGPT are two well-liked generating tools.
- AI can assist e-commerce companies in optimizing inventory management and supply chain operations. AI-powered systems can predict demand by analyzing prior sales data and adjusting inventory levels accordingly. This can assist firms in

reducing the risk of overstocking or understocking, as well as improving overall supply chain efficiency.

6 Challenges and Considerations

While AI-powered e-commerce offers several benefits, there are also some challenges and considerations that businesses should keep in mind. Here are some of the key challenges and considerations:

- In order to provide precise forecasts and suggestions, artificial intelligence systems need enormous volumes of high-quality data. Gathering, purifying, and guaranteeing data privacy compliance are common challenges for e-commerce businesses.
- Privacy and security are major issues when handling consumer data. There are serious repercussions for any data breaches, thus it is imperative to ensure compliance with laws like the CCPA and GDPR.
- It can be costly to develop and execute AI solutions, particularly for smaller e-commerce companies. Expenses include buying technology, developing or acquiring AI models, and employing data scientists and engineers. Be not misled: there are AI systems that provide a free trial to experience the benefits.
- AI and machine learning skill is in limited supply. Competition for top talent is intense, making it difficult to find and keep qualified data scientists and engineers.
- A complicated world of AI ethics and laws must be negotiated by e-commerce businesses. To prevent legal issues and damage to one's reputation, adherence to different local, state, and federal laws is crucial.

Businesses must be mindful of the challenges and considerations when using AI in e-commerce, including data privacy and security, ethical concerns, human oversight, training data, and customer acceptance. In spite of these challenges, using AI to e-commerce presents significant chances for companies to improve consumer experiences, streamline processes, and get a competitive advantage. Ecommerce businesses that want to fully utilize AI in their operations must successfully handle these obstacles.

7 Future of AI in E-Commerce

In the upcoming years, one of the industry's most likely to benefit from the use of AI-based solutions is customer service. Chatbots and virtual assistants that automate customer service in e-commerce have the biggest advantages when it comes to speed and accuracy of responses and improving the customer experience (i.e., consistently giving them the feeling of "being taken care of"—regardless of the time of day or night).

Sentiment analysis, natural language processing, and predictive analytics are just a few of the advances in artificial intelligence (AI) that are transforming the electronics industry [19]. In the rapidly evolving field of e-commerce, artificial intelligence is beginning to revolutionize business operations. Now let's examine how some of the largest online merchants, like Amazon, Alibaba, eBay, and others, are using AI to build their brands and advance their companies [20].

The majority of e-commerce advancements in the future, as you may already be aware, hinge on the wider application of artificial intelligence. e-commerce services now employ artificial intelligence, but more intelligent robots will drive the industry's growth in the future. In order to obtain a competitive advantage in the e-commerce industry, Alibaba, a major participant, has adopted AI helpers like Tmall Genie and Ali Assistant. 95 percent of written and spoken client questions are handled by their AI-powered customer support chatbot. Alibaba also uses AI for intelligent product suggestions, improved search functionality, and effective delivery route mapping.

One of the most promising areas for AI in e-commerce is predictive analytics. AI-powered algorithms can analyze vast amounts of data, such as customer behavior, purchase history, and external factors like weather and social trends, to generate accurate predictions of future sales and demand. This will enable online retailers to optimize their inventory, pricing, and marketing strategies to maximize revenue and minimize waste.

Personalization is another key area where AI can transform e-commerce. By using machine learning algorithms to analyze customer data and behavior, online retailers can provide personalized recommendations, promotions, and content to each individual customer, improving engagement and loyalty.

Finally, fraud detection and prevention are an area where AI is already making a significant impact. Machine learning algorithms can analyze vast amounts of data to identify fraudulent transactions and prevent chargebacks, protecting online retailers and their customers from financial losses. AI has already transformed the e-commerce industry, and its potential to continue to evolve and transform the sector in the coming years is enormous. Online retailers that embrace AI technology and use it to enhance the customer experience, optimize their operations, and increase revenue will be the ones that thrive in the future.

8 Conclusion

AI is utilized in marketing for e-commerce companies to better understand their clientele and spot emerging trends and patterns in consumer behavior. More focused offerings, marketing efforts, and advertisements may also be made by businesses thanks to it. To increase the volume of content produced and make sure that it speaks to the target market for the brand, marketers employ generative AI technologies. Finally, marketers employ Omni channel strategies and AI to retarget prospective consumers in an effort to persuade them to buy products and services.

AI in e-commerce will benefit businesses greatly if it is used more widely. only because it offers answers that are quicker, less expensive, more effective, and less prone to human error. Our world is becoming increasingly digitalized by the day, thus the idea of e-commerce, which makes our lives easier every day, needs to change to meet the needs of consumers and keep up with the latest advancements in technology.

Although AI is currently present in many forms of online commerce, machine learning and AI will ultimately change the e-commerce sector in the coming years. AI has a significant impact on the e-commerce sector by bringing new ideas and improved consumer experiences. Personalized shopping experiences, chatbots, virtual assistants, voice search, and product suggestions are a few of the most notable applications of AI in e-commerce. These are not the only advantages. AI has the potential to augment the benefits of gathering and analysing client data by providing customized online retailing services based on individual customer preferences and interests.

AI and data collected from consumers and businesses are used by today's e-commerce companies to predict future results and adjust marketing campaigns more successfully, allowing them to make well-informed decisions. Finally, enterprises need to adopt all e-commerce concepts, such as customer journey, storage, shipping, secure payment systems, and so on, in order to advance artificial intelligence technologies and remain one step ahead of the competition in the big data era.

References

1. Di Vaio, A., Palladino, R., Hassan, R., & Escobar, O. (2020). Artificial intelligence and business models in the sustainable development goals perspective: A systematic literature review. *Journal of Business Research, 121*, 283–314.
2. Duan, Y., Edwards, J. S., & Dwivedi, Y. K. (2019). Artificial intelligence for decision making in the era of big data–evolution, challenges and research agenda. *International Journal of Information Management, 48*, 63–71.
3. Kalakota, R., & Whinston, A. B. (1997). Electronic commerce: a manager's guide. Addison-Wesley Professional.
4. Holsapple, C. W., & Singh, M. (2000). Electronic commerce: From a definitional taxonomy toward a knowledge-management view. *Journal of Organizational computing and Electronic Commerce, 10*(3), 149–170.
5. Tan, F. T. C., Pan, S. L., & Zuo, M. (2019). Realising platform operational agility through information technology–enabled capabilities: A resource-interdependence perspective. *Information Systems Journal, 29*(3), 582–608.
6. Gielens, K., & Steenkamp, J. B. E. (2019). Branding in the era of digital (dis) intermediation. *International Journal of Research in Marketing, 36*(3), 367–384.
7. Hamad, H., Elbeltagi, I., & El-Gohary, H. (2018). An empirical investigation of business-to-business e-commerce adoption and its impact on SMEs competitive advantage: The case of Egyptian manufacturing SMEs. *Strategic Change, 27*(3), 209–229.
8. Kaplan, A., & Haenlein, M. (2019). Siri, Siri, in my hand: Who's the fairest in the land? On the interpretations, illustrations, and implications of artificial intelligence. *Business Horizons, 62*(1), 15–25.

9. Kumar, N., Venugopal, D., Qiu, L., & Kumar, S. (2019). Detecting anomalous online reviewers: An unsupervised approach using mixture models. *Journal of Management Information Systems, 36*(4), 1313–1346.
10. Deng, S., Tan, C. W., Wang, W., & Pan, Y. (2019). Smart generation system of personalized advertising copy and its application to advertising practice and research. *Journal of Advertising, 48*(4), 356–365.
11. Kietzmann, J., Paschen, J., & Treen, E. (2018). Artificial intelligence in advertising: How marketers can leverage artificial intelligence along the consumer journey. *Journal of Advertising Research, 58*(3), 263–267.
12. Kumar, T., & Trakru, M. (2020). The colossal impact of artificial intelligence. E-commerce: Statistics and facts. *Int. Res. J. Eng. Technol. (IRJET), 6*, 570–572.
13. Soni, N., Sharma, E. K., Singh, N., & Kapoor, A. (2019). Impact of artificial intelligence on businesses: From research, innovation, market deployment to future shifts in business models. arXiv preprint arXiv:1905.02092.
14. Soni, N., Sharma, E. K., Singh, N., & Kapoor, A. (2020). Artificial intelligence in business: From research and innovation to market deployment. *Procedia Computer Science, 167*, 2200–2210.
15. Krishnan, C., Gupta, A., Gupta, A., & Singh, G. (2022). Impact of artificial intelligence-based chatbots on customer engagement and business growth. *Deep learning for social media data analytics* (pp. 195–210). Springer International Publishing.
16. Urbanke, P. (2016). Essays on Predictive Analytics in E-Commerce (Doctoral dissertation, Niedersächsische Staats-und Universitätsbibliothek Göttingen).
17. Morsi, S. (2020). A Predictive analytics model for e-commerce sales transactions to support decision making: A case study. *International Journal of Computer and Information Technology* (2279–0764), *9*(1).
18. Awan, N. N. (2023). How predictive analytics can empower your decision making. *Handbook of Big Data Research Methods*, 117.
19. Moorthi, K., Dhiman, G., Arulprakash, P., Suresh, C., & Srihari, K. (2021). Withdrawn: A survey on impact of data analytics techniques in E-commerce.
20. Ahmed, A. A. A., Agarwal, S., Kurniawan, I. G. A., Anantadjaya, S. P., & Krishnan, C. (2022). Business boosting through sentiment analysis using artificial intelligence approach. *International Journal of System Assurance Engineering and Management, 13*(Suppl 1), 699–709.

Impact of Artificial Intelligence on Purchase Intention: A Bibliometric Analysis

Aatam Parkash Sharma, Naresh Kumar Sharma, Neeru Sidana, and Richa Goel

Abstract Technology development has brought about new business practices and improved its ability to affect consumer behavior. Technology use also makes it easier to comprehend customer needs and provide better services to them. Therefore, in today's technology-driven marketing scenarios, it is crucial to comprehend the development of artificial intelligence (AI) and how it affects consumers' purchase intentions. As artificial intelligence (AI) or AI-enabled services become more prevalent in the digital marketplace, we are seeing an impact on consumer purchase intentions. With the aid of bibliometric analysis, this paper seeks to review research on artificial intelligence and how it affects consumer purchase intention. The Web of Science and Scopus databases were used to gather the sample size for the study, which contained 85 papers from the years 2005 through 2023. The study reviews the annual scientific production along with the most prolific authors, articles and journals. Additionally, this study looks at the co-occurrence analysis to identify the thematic clusters utilizing the author's keywords. The annual scientific production analysis reveals a 12.25% yearly growth rate for research publications in the field under study. The top two highly cited journals in the field are the International Journal of Information Management and Telematics and Informatic with 320 and 97 citations respectively. Furthermore, the thematic clusters derived from the co-occurrence network highlight highly used keywords such as artificial intelligence, purchase intention, e-commerce and consumer behavior.

Keywords Artificial intelligence · Purchase intention · Consumer behavior · Bibliometric analysis

A. P. Sharma · N. K. Sharma
Department of Marketing and Supply Chain Management, School of Business Studies, Central University of Jammu, Jammu and Kashmir, India

N. Sidana (✉)
Amity School of Economics, Amity University, Noida, India
e-mail: nsidana@amity.edu

R. Goel
Department of Economics and IB, SCMS Noida, Symbiosis International University, Pune, India

L. Gaur and A. Abraham (eds.), *Role of Explainable Artificial Intelligence in E-Commerce*, Studies in Computational Intelligence 1094,
https://doi.org/10.1007/978-3-031-55615-9_5

1 Introduction

The involvement of technology in B2C (business-to-consumer) interaction has changed the current patterns of consumer engagement vis-à-vis various retailing activities [1, 2]. The consumption pattern of consumers has changed from shops to online platforms [2]. Due to the expansion of online retailing and the availability of big data, various brands have started using artificial intelligence (AI)-based marketing strategies for better consumer engagement, strategic marketing planning and market success, along with technical AI algorithms to solve various marketing related problems [3, 4]. Artificial intelligence (AI) is a technical aid that helps consumers to differentiate between products according to their needs and interest. Online platforms provide proactive product recommendations to the consumer according to their search with the help of digital assistants or artificial intelligence [1, 4, 5]. Nowadays, consumers are becoming more dependent on artificial intelligence for seeking information regarding products. If the information provided by the human service agent of any retailing website does not satisfy their desire they become frustrated and may move to another competitor retailer [6]. This is why, AI technology gained importance in online retailing.

Digital platforms like social media and e-commerce websites provide a digital marketplace for companies to do their business [7]. In the current scenario, artificial intelligence is involved in operational marketing that includes the identification of risk, targeting of consumers, brand advertising and pricing to maximize the profit [8]. Hence, AI is substantially changing brand preferences, marketing strategies and consumer attitudes. Furthermore, AI is applied in different practices of marketing like segmentation, personalization, pricing and sales forecasting [9]. It can be witnessed that, these impending shifts in online retailing have a considerable impact on consumer purchase intention [7]. However, many industrial practitioners are still unaware of the role of AI in purchase intention.

Therefore, it presents a gap that is very important and timely to understand how artificial intelligence is impacting purchase intention. In addition, there are only a few pieces of research evidence which are addressing the issue. The study tried to fulfill the following objectives:

1. To explore the present state of research in the field of AI and purchase intention.
2. To find out the top contributing authors, articles and journals in the area of AI and purchase intention.
3. To find out the major keywords associated with AI and purchase intention.

With the help of bibliometric analysis, the current study addresses the above research objectives. The bibliometric analysis examines a complete set of research in a specific area from a quantitative perspective [7, 10]. In this article, the authors have designed a structured analysis of Artificial intelligence and its relationship with purchase intention.

2 Conceptual Background

The shreds of AI can be discovered during the 1950s when English polymath Alan Turing started a test to verify if the machine can perform a function like humans [7]. The term AI can be defined as the science of training machines that can perform like humans by using big data [11, 12]. It is all about machines doing things that require human intelligence, understanding languages, reasoning and exploring the physical world [13].

Social media and e-commerce websites are generally used by consumers to seek information regarding products or services whereas previous consumer feedback is used to make any purchase decision [14, 15, 16]. Purchase intention can be defined as "the possibility that consumer will plan or be willing to purchase a certain product or service in the future" [17]. It can also be considered as one previous step than actual buying [18]. In other words, purchase intention is dependent on the consumer's willingness to purchase [16].

Artificial intelligence has been widely used in e-commerce websites and online business service fields. The consumer gets more satisfaction from AI-based services than human-based services [19]. Moreover, these human-like services provided by artificial intelligence influence consumer purchase intention, which is why online retailers use these kinds of AI services more in their websites to stimulate consumer purchase intention [1, 20]. Artificial intelligence is redesigning consumer interaction while online shopping and influencing their purchase intention [2].

3 Methodology

The study uses bibliometric analysis tools to examine the underlying research objectives. The bibliometric analysis is defined as the application of quantitative tools to bibliographic data [21]. A bibliometric analysis is an established and reliable method [22]. With the help of bibliometric analysis, a systematic, objective and transparent analysis of the present literature can be performed [23]. Bibliometric analysis generally contains two parts: performance analysis and science mapping.

In the initial phase, this study followed a simple two-step process to screen and select the appropriate study. Firstly, two bibliographic databases the Web of Science and Scopus were used to select the research sample. The following research string [("artificial intelligence" AND" purchase intention")] is used in both databases: For the Scopus database, the string is as follows: *Title-Abs-Key ("Artificial Intelligence" And "Purchase Intention") And (Limit-To (Pubstage, "Final")) And (Limit-To (Doctype, "Ar")) And (Limit-To (Language, "English"))* and for Web of Science the string is as follows: *"artificial intelligence" AND "purchase intention" (All Fields) and Article (Document Types) and English (Languages).* The data was extracted on 11th April 2023.

This study accommodates papers from 2005 to 2023. Further, the string includes only papers in the English language and peer-reviewed articles only. The study excluded papers other than English language and documents type as book chapters, editorials, conference abstracts, discussions, etc. The multidisciplinary research subject areas have been explored as the relationship between artificial intelligence and purchase intention has been analyzed in varied scopes. A total no of 150 articles were found in both databases. After removing duplicate articles, only 85 were selected to run the analysis on.

In this study, a mixture of both performance analysis and science mapping is performed to accommodate the research objective.

4 Findings

The bibliometric study of Artificial Intelligence and its relation with Purchase Intention is presented in this part. This analysis is carried out in such a way that it answers the study objectives. The bibliometrix package from the R software was utilized for the analysis, which is common for this sort of research.

4.1 Annual Scientific Production

The annual scientific production analysis is used to explore the present state of research (Research Objective 1). The yearly publishing activity pattern is favorable for examining the knowledge about the introductory, growth and maturity phases of any given research topic. Fig. 1 is presenting the research articles produced annually from the year 2005 to March 2023. In the initial phase, from 2005 to 2014 there was a very slow increase in the research publication exploring the relationship between AI and Purchase intention. However, from the period of 2014 to 2018 we can witness a gradual increase in the research output. After 2018, the total number of publications started growing rapidly. Furthermore, in 2022, the total number of articles has almost doubled that of the previous year 2021. It shows that more researchers have started working on this research area. The steep rise in the number of articles also highlights that organizations have started using artificial intelligence more for knowing consumer intentions which makes this area more important to explore. For instance, Yen and Chiang [24] highlights that organizations are using Artificial intelligence like chatbots for meeting their consumer expectations in a better way. Researchers have suggested that businesses are using AI in their strategic marketing model to better understand and meet consumer purchase intention [25]. With the growing e-commerce platform, businesses have adopted various AI-enabled aids for a better consumer experience. Furthermore, researchers are keenly exploring this field of study since 2018, mainly because organizations have started adopting AI in their business models more avidly from that time period [26].

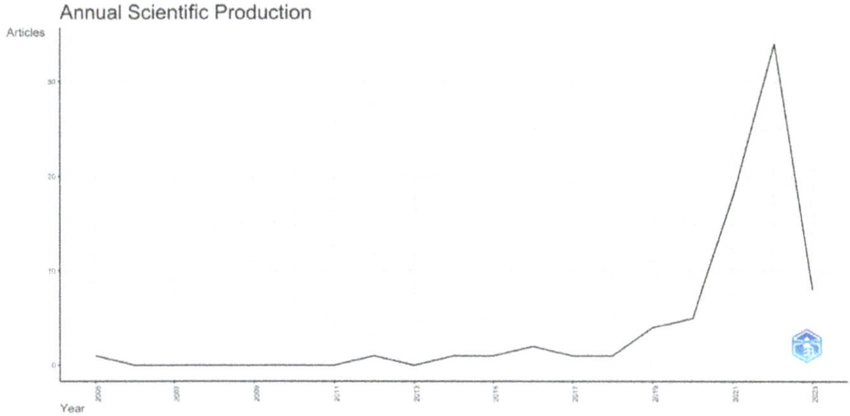

Fig. 1 The annual publication output of articles included in this review

The annual scientific production analysis reveals a 12.25% yearly growth rate for research publications in the field under study. A total of 8 articles is published in the first quarter of 2023 which also forecast that many more can be done.

4.2 Analysis of the Authors, Articles and Journals

Figure 2 shows the result of the top 10 most productive authors in the field of Artificial Intelligence and purchase intention indexed by the bibliomatrix package in R software. Yuangao Chen stands in the first position with a total no of 5 articles followed by Ying Liu, Yogesh Kumar Dwivedi, Cheng-Chung Chen and Ohbyung Kwon. The most cited author among the top 10 authors is Yogesh Kumar Dwidevi with a total of 4 articles and 384 citations followed by Ohbyung Kwon with 3 articles and 135 citations (Table 1).

With the help of most globally cited articles in biblioshiny, the top 10 most influential articles were analyzed on the bases of their citation score (Table 2). Moreover, Table 2 also shows the name of journals and the number of citations they have received. "*Setting the future of digital and social media marketing research: Perspectives and research propositions*" by Yogesh Kumar Dewvedi and others have earned 320 citations which is the highest among others. Followed by Kwonsang Sohn and Ohbyung Kwon, R. Pillai, et al., and CammuCrolic et al., with 97, 82 and 59 citations respectively (to the date of performing the study). The higher number of citations shows the quality of work done in the area of Artificial Intelligence and Purchase Intention. The growing number of citations also shows the importance of the research area.

The bibliometric analysis identifies the main journals and their contribution to the extant literature on this particular research area (Xu et al., 2018). Table 3 demonstrates

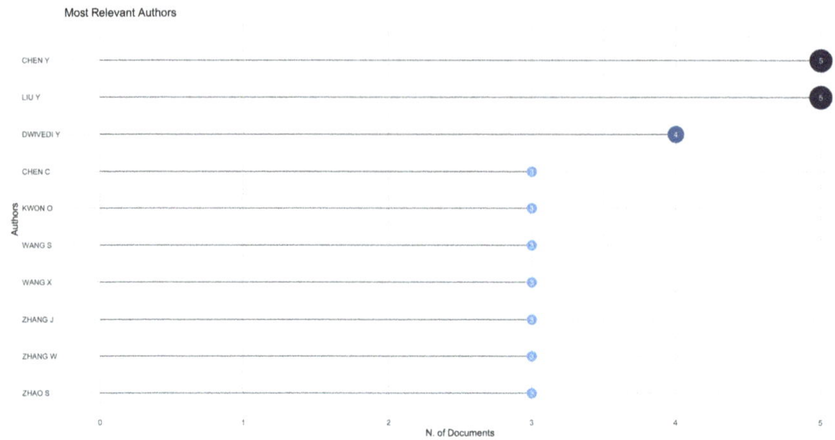

Fig. 2 Most prolific authors in the field of AI and purchase intention as per articles

Table 1 Top 10 contributing authors

Authors	Articles	Total citation	Average citation/article
Yuangao Chen	5	72	14.4
Yu Liu	5	5	1
Yogesh K. Dwivedi	4	384	96
Chia Chen	3	8	2.7
Obhyung Kwon	3	135	45
WANG S	3	21	7
Xinping Wang	3	20	6.7
Jianya Zhang	3	49	16.3
Wei Zhang	3	7	2.3
Shiyue Zhao	3	5	1.7

the journals that published the most articles on the relationship between Artificial intelligence and Purchase Intention. According to Table 3, the highest-publishing journal in this field is *Sustainability,* which has contributed 4 articles on this topic. The articles in this journal have received 43 citations. It is followed by the *Journal of Retailing and Consumer Services, International Journal of Information Management, The Journal of Business Research and Technological Forecasting and Social Change.* On the basis of citation score International Journal of Information Management has received the highest citation of 350 with 3 articles. Moreover, the top 5 journals based on publication are mentioned in Table 3.

Table 2 Top 10 most influential articles

Rank	Article	Authors	Journal	Total citation
1	Setting the future of digital and social media marketing research: Perspectives and research propositions	Joyesh K. Dewvedi et al	International Journal of Information Management	320
2	Technology acceptance theories and factors influencing artificial Intelligence-based intelligent products	Kwonsang Sohn and Ohbyung Kwon	Telematics and Informatics	97
3	Shopping intention at AI-powered automated retail stores (AIPARS)	R. Pillai, B. Sivathanu and Joyesh K. Dewvedi	Journal of Retailing and Consumer Services	82
4	Blame the Bot: Anthropomorphism and Anger in Customer–Chatbot Interactions	Cammy Crolic et al	Journal of Marketing	59
5	Alexa, do voice assistants influence consumer brand engagement?—Examining the role of AI powered voice assistants in influencing consumer brand engagement	Graeme McLean, Kofi Osei-Frimpong and Jennifer Barhorst	Journal of Business Research	56
6	Factors impacting consumers' intention toward adoption of electric vehicles in Malaysia	Shahla Asadi et al	Journal of Cleaner Production	56
7	Conversational commerce: entering the next stage of AI-powered digital assistants	Janarthanan Balakrishnan and Yogesh K. Dwivedi	Annals of Operations Research	45
8	Predicting movie Box-office revenue by exploiting large-scale social media content	Ting Liu et al	Multimedia Tools and Applications	44
9	Trust me, if you can: a study on the factors that influence consumers' purchase intention triggered by chatbots based on brain image evidence and self-reported assessments	Chiahui Yen and Ming-Chang Chiang	Behaviour and Information Technology	43
10	Competitive intelligence in social media Twitter: iPhone 6 versus Galaxy S5	Yoosin Kim et al	Online Information Review	43

Table 3 Top 5 Most influential journals

S. No.	Journal	No. of articles	Total citation
1	Sustainability	4	43
2	Journal of Retailing and Consumer Services	3	105
3	International Journal of Information Management	2	350
4	Journal of Business Research	2	57
5	Technological Forecasting and Social Change	2	21

Table 4 Highest occurring keywords in each cluster

Cluster #1			
Keyword	Total Occurrence	Betweenness	Closeness
Purchase intention	20	27.28	0.06
e-commerce	5	0	0.05
Chatbots	3	0	0.05
ai chatbot	2	0	0.03
Perceived hedonic value	2	0	0.03
Trust	2	0	0.03
Cluster #2			
Keyword	Total Occurrence	Betweenness	Closeness
Artificial intelligence	23	40.71	0.07
Perceived risk	3	0	0.04
Anthropomorphism	2	0	0.04
Consumer behavior	2	0	0.04
Machine learning	2	0	0.04
Social media marketing	2	0	0.04

4.3 Top Keywords Clusters and Connection

Table 4 presents the summaries of the cluster connection and the top keywords in each cluster, respectively. The central theme of each cluster is identified on the basis of keywords' total occurrence (TO), betweenness and closeness in the cluster.

4.3.1 Cluster #1: Purchase Intention

The first cluster shows a total of 6 nodes where purchase intention has the highest occurrence of 20 times followed by e-commerce, chatbots, ai-chatbot, perceived hedonic value and trust. The term purchase intention is the main node in this cluster with the betweenness and closeness of 27.28 and 0.06. Various technology acceptance models are utilized to understand the relationship between purchase intention and

Fig. 3 Co-occurrence of keyword distribution in the reviewed articles

artificial intelligence where the Value-based adoption model (VAM) performed the best in modeling. Enjoyment is the factor that influences the most consumer purchase intention [27]. Chatbots allow businesses to enhance the consumer experience and meet their expectation through real-time interaction in e-commerce platforms [24]. Studies have also found that the accuracy, insight and interaction experience in artificial intelligence positively impact the perceived utility value and hedonic values [28]. E-commerce businesses are converting consumers' perceived service quality into purchase intention with the help of Ai-chatbots [19] (Fig. 3).

4.3.2 Cluster #2: Artificial Intelligence

Cluster #2 contains 5 keywords out of which artificial intelligence has the highest number of occurrences (23). Artificial Intelligence is the main node in this cluster containing a betweenness of 40.71 and closeness of 0.07. Immersive technological changes in retail services have changed the consumption environment, where the quality of AI positively influences the consumer's behavior. The quality of AI technology opens a new avenue to consumer engagement [29]. Social media marketing and online shopping platforms are using AI technology to provide more accurate and personalized services experience for consumers [28, 30]. To overcome the consumer's perceived risk and trust in online shopping, businesses use AI technology [31, 32]. This highlights, how artificial intelligence is related to all the keywords in the cluster.

5 Conclusion

The present study has made an attempt to understand the current state and ambit of research in the field of Artificial Intelligence concerning Purchase Intention by using bibliometric analysis. Articles from Web of Science and Scopus were reviewed ranging from the period 2005 to 2023 and the annual scientific production analysis reveals the annual growth rate for research publication in the tune of 12.25%. The results reveal that the most cited author is Yogesh Kumar Dwidevi with a total of 4 articles and 384 citations followed by Ohbyung Kwon with 3 articles and 135 citations. The highest number of articles publishing in the area is in the *Sustainability* journal but International Journal of Information Management has received the highest citation of 350 with 3 articles. Further, the study reveals that purchase intention is the main node in Cluster-1 with the betweenness and closeness of 27.28 and 0.06. Cluster -2 contains 5 keywords out of which artificial intelligence has the highest number of occurrences (23). The growing number of articles and citations reveals that more and more researchers are attracted to the emerging area as this can influence the purchasing behavior of consumers by earning loyalty for the products and services they are offering.

6 Future Implication

The present study shows that the research in the area of Artificial Intelligence has picked up speed (arose) dramatically in the past few years and most of the articles are coming from the countries like China, India, South Korea and the USA. This study has used the bibliometrix R package for the analysis, other may use different software like VOS viewer and Hitsite. The study was also limited to two databases Web of Science and Scopus, further studies can consider additional databases for research work. Though the present study is in a sense comprehensive, reliable and valid yet it was carried out with limitations. The future research needs to extend to other variables so that a macro- perspective could be evolved with regard to the adoption of Artificial intelligence in influencing purchasing patterns of customers.

References

1. Malhotra, G., & Ramalingam, M. (2023). Perceived anthropomorphism and purchase intention using artificial intelligence technology: Examining the moderated effect of trust. *Journal of Enterprise Information Management*. https://doi.org/10.1108/JEIM-09-2022-0316.
2. Zogaj, A, Mähner, P. M., Yang, L., Tscheulin, D. K. (2023, Jan). It's a Match! The effects of chatbot anthropomorphization and chatbot gender on consumer behavior. *Journal of Business Search*, *155*. https://doi.org/10.1016/j.jbusres.2022.113412.
3. Das, G., Jain, S. P., Maheswaran, D., Slotegraaf, R. J., & Srinivasan, R. (2021). Pandemics and marketing: Insights, impacts, and research opportunities. *Journal of the Academy of Marketing Science*, pp. 835–854. https://doi.org/10.1007/s11747-021-00786-y/Published.
4. Bawack, R. E., Wamba, S. F., & Carillo, K. D. A. (2021, Feb). A framework for understanding artificial intelligence research: Insights from practice. *Journal of Enterprise Information Management*, *34*(2), 645–678. https://doi.org/10.1108/JEIM-07-2020-0284.
5. Balakrishnan, J., & Dwivedi, Y. K. (2021). Conversational commerce: Entering the next stage of AI-powered digital assistants. *Ann Oper Res*. https://doi.org/10.1007/s10479-021-04049-5.
6. Troshani, I., Hill, S. R., Sherman, C., & Arthur, D. (2021). Do we trust in AI? Role of anthropomorphism and intelligence. *Journal of Computer Information Systems*, *61*(5), 481–491. https://doi.org/10.1080/08874417.2020.1788473
7. Varsha, P. S., Akter, S., Kumar A., Gochhait, S., & Patagundi, B.. (2021, Jul 01). The impact of artificial intelligence on branding: A bibliometric analysis (1982–2019). *Journal of Global Information Management*, *29*(4), 221–246, IGI Global. https://doi.org/10.4018/JGIM.20210701.oa10.

8. Marinchak, C. M. D., Forrest, E., & Hoanca, B. (2018, Jan). Artificial intelligence: Redefining marketing management and the customer experience. *8*(2), 14–24. https://doi.org/10.4018/IJEEI.2018070102.
9. Columbus, L. 10 Charts that will change your perspective of AI In marketing. https://www.forbes.com/sites/louiscolumbus/2019/07/07/10-charts-that-will-change-your-perspective-of-ai-in-marketing/amp/. Accessed: 06 May 2023. [Online]. Available: https://www.forbes.com/sites/louiscolumbus/2019/07/07/10-charts-that-will-change-your-perspective-of-ai-in-marketing/amp/.
10. Donthu, N., Kumar, S., Mukherjee, D., Pandey, N., & Lim, W. M. (2021). How to conduct a bibliometric analysis: An overview and guidelines. *Journal of Business Research, 133*, 285–296. https://doi.org/10.1016/j.jbusres.2021.04.070
11. Davenport, T., Guha, A., Grewal, D., & Bressgott, T. (2020). How artificial intelligence will change the future of marketing. *Journal of the Academy of Marketing Science, 48*(1), 24–42. https://doi.org/10.1007/s11747-019-00696-0
12. Kumar, V., Ramachandran, D., & Kumar, B. (2021). Influence of new-age technologies on marketing: A research agenda. *Journal of Business Research, 125*, 864–877. https://doi.org/10.1016/j.jbusres.2020.01.007
13. Huang, M. H., & Rust, R. T. (2018). Artificial intelligence in service. *Journal of Service Research, 21*(2), 155–172. https://doi.org/10.1177/1094670517752459
14. Brown, M., Pope, N., & Voges, K. (2003). Buying or browsing?: An exploration of shopping orientations and online purchase intention. *European Journal of Marketing, 37*(11–12), 1666–1684. https://doi.org/10.1108/03090560310495401
15. Racherla, P., & Friske, W. (2012). Perceived "usefulness" of online consumer reviews: An exploratory investigation across three services categories. *Electronic Commerce Research and Applications, 11*(6), 548–559. https://doi.org/10.1016/j.elerap.2012.06.003
16. Bahtar, A. Z., & Muda, M. (2016). The impact of user—generated content (UGC) on product reviews towards online purchasing—A conceptual framework. *Procedia Economics and Finance, 37*, 337–342. https://doi.org/10.1016/s2212-5671(16)30134-4
17. Wu, P. C. S., Yeh, G. Y. Y., & Hsiao, C. R. (2011). The effect of store image and service quality on brand image and purchase intention for private label brands. *Australasian Marketing Journal, 19*(1), 30–39. https://doi.org/10.1016/j.ausmj.2010.11.001
18. De Magistris, T., & Gracia, A. (2008). The decision to buy organic food products in Southern Italy. *British Food Journal, 110*(9), 929–947. https://doi.org/10.1108/00070700810900620
19. Qin, M., Zhu, W., Zhao, S., & Zhao, Y. (2022, April). Is artificial intelligence better than manpower? The effects of different types of online customer services on customer purchase intentions. *Sustainability (Switzerland), 14*(7). https://doi.org/10.3390/su14073974.
20. Bartneck, C., Kulić, D., Croft, E., & Zoghbi, S. (2009). Measurement instruments for the anthropomorphism, animacy, likeability, perceived intelligence, and perceived safety of robots. *International Journal of Social Robotics, 1*(1), 71–81, Kluwer Academic Publishers. https://doi.org/10.1007/s12369-008-0001-3.
21. Donthu, N., Kumar, S., Pandey, N., Pandey, N., & Mishra, A. (2021). Mapping the electronic word-of-mouth (eWOM) research: A systematic review and bibliometric analysis. *Journal of Business Research, 135*, 758–773. https://doi.org/10.1016/j.jbusres.2021.07.015
22. Hood, W. W., & Wilson, C. S. (2001). The literature of bibliometrics, scientometrics, and informetrics. Kluwer Academic Publishers.
23. Rivai, F. A., Navimipour, N. J., & Yalcın, S. (2023). Multimedia big data computing mechanisms: A bibliometric analysis. *Multimed Tools Appl, 82*(2), 2765–2781. https://doi.org/10.1007/s11042-022-12988-9
24. Yen, C., & Chiang, M. C. (2021). Trust me, if you can: A study on the factors that influence consumers' purchase intention triggered by chatbots based on brain image evidence and self-reported assessments. *Behaviour and Information Technology, 40*(11), 1177–1194. https://doi.org/10.1080/0144929X.2020.1743362
25. Jiang, K., Qin, M., & Li, S. (2022). Chatbots in retail: How do they affect the continued use and purchase intentions of Chinese consumers? *Journal of Consumer Behaviour, 21*(4), 756–772. https://doi.org/10.1002/cb.2034

26. Thormundsson, B. Total global AI investment 2015–2021 | Statista'. Accessed: 09 May 2023. [Online]. Available: https://www.statista.com/statistics/941137/ai-investment-and-funding-worldwide/.
27. Sohn, K., & Kwon, O. (2020). Technology acceptance theories and factors influencing artificial Intelligence-based intelligent products. *Telematics and Informatics*, *47*. https://doi.org/10.1016/j.tele.2019.101324.
28. Yin, J., Qiu, X. (2021, May). Ai technology and online purchase intention: Structural equation model based on perceived value. *Sustainability (Switzerland)*, *13*(10). https://doi.org/10.3390/su13105671.
29. Dwivedi, Y. K., et al. (2022). Metaverse beyond the hype: Multidisciplinary perspectives on emerging challenges, opportunities, and agenda for research, practice and policy. *International Journal of Information Management, 66*, 102542. https://doi.org/10.1016/J.IJINFOMGT.2022.102542
30. Simay, A. E., Wei, Y., Gyulavári, T., Syahrivar, J., Gaczek, P., & Hofmeister-Tóth, Á. (2022). The e-WOM intention of artificial intelligence (AI) color cosmetics among Chinese social media influencers. *Asia Pacific Journal of Marketing and Logistics*, vol. ahead-of-print, no. ahead-of-print. https://doi.org/10.1108/APJML-04-2022-0352/FULL/XML.
31. Li, X., Yang, J., Wang, X., & Lei, D. (2012). The impact of country-of-origin image, consumer ethnocentrism and animosity on purchase intention. *Journal of Software, 7*(10), 2263–2268. https://doi.org/10.4304/jsw.7.10.2263-2268
32. Habib, S., & Hamadneh, N. N. (2021). Impact of perceived risk on consumers technology acceptance in online grocery adoption amid covid-19 pandemic. *Sustainability (Switzerland), 13*(18). https://doi.org/10.3390/su131810221.

Chatbot-XAI—The New Age Artificial Intelligence Communication Tool for E-Commerce

Kavita Thapliyal and Manjul Thapliyal

Abstract XAI chatbots in e-commerce refer to chatbots that are customized to use natural language processing (NLP) and machine learning (ML) algorithms to recognise and comprehend customer queries and deliver personalized and tailored recommendations and assistance to customers during online shopping. The primary benefit of XAI chatbots in ecommerce is that they can support and clarify their decision-making process and procedures to customers more clearly and helps in building trust and transparency, by decreasing the possibility for errors or biases. Today XAI chatbot can explain why it recommends and proposes a particular product to his customer based on the prior preferences, previous purchases, and browsing history. This chapter aims to provide an overview of how XAI chatbots can be used as a tool in ecommerce to improve customer experience and increase sales. Overall, XAI chatbots have the potential to revolutionize the ecommerce industry however, they must be designed and implemented carefully to ensure they are ethical, secure, and compliant with privacy regulations. The chapter will in detail delve into the technical aspects of XAI chatbots, including the machine learning algorithms and natural language processing techniques used to build them. With the help of detailed pictorial representation, it will exhibit the importance of transparency and interpretability in XAI chatbots with various techniques and approaches of explaining the chatbot's decision-making process to customers. Overall, this chapter will provide a comprehensive overview of XAI chatbots as a tool in ecommerce and their potential impact on the industry.

Keywords XAI (Explainable Artificial Intelligence) · Chatbots · Ecommerce · Natural language processing (NLP) · Machine learning (ML) · Personalized recommendation · Customer Support · Black box · White box · Attention maps · Algorithms · Decision tree · Random FOREST · Linear and logistic regression

---

K. Thapliyal (✉)
Amity International Business School (AIBS), Amity University, Noida, India
e-mail: kthapliyal@amity.edu

M. Thapliyal
Visions Ahead Foundation, Noida, India
e-mail: visionsaheadfdn@gmail.com

L. Gaur and A. Abraham (eds.), *Role of Explainable Artificial Intelligence in E-Commerce*, Studies in Computational Intelligence 1094,
https://doi.org/10.1007/978-3-031-55615-9_6

1 Introduction

XAI, is the latest AI model tool based on digital systems providing clear, concise, and easy to interpret explanations of its actions and results [1]. Using XAI systems, conversational Chatbots which are powered by AI can gain immensely from their superior techniques in diversified fields, like ecommerce.

XAI enabled chatbots used in ecommerce have a major advantage in providing clear and personalized suggestions to consumers based on their past data and choice preferences [2]. Users can understand and judge the recommendation of the chatbot based on its explanations and logical data. This leads to better trust, and more engagement with higher conversion ratio. Moreover, chatbots powered by XAI can easily identify and erase bias factors from their recommendation tools, thus making ecommerce more inclusive and transparent [3]. This also eliminates discrimination against a particular user or consumer groups for that matter and ensures no one loses on any benefits provided for all. This enhances the overall consumer experience, trust and buyer satisfaction when enabled by recommendations based on transparency, logic, and personalized touch.

Chatbots are user friendly agents designed through computer programming powered by Artificial Intelligence to stimulate interesting conversations with humans by using Machine Learning (ML) and Natural Language Processing (NLP) digital techniques [4]. XAI is an advanced form of AI which makes it more user friendly through its intricate models and systems by focusing on its easy understandability and transparency towards its users. This helps in improving the user experience and enhancing trust factors of consumers [5]. Knowledge Graphs (KG) are used across various domains through semantic models to outline properties, relationships, and classes of different entities. This helps AI in better interpretation and explainable.

XAI chatbots has the potential to transform the ecommerce sector by improving the customer communicative experience, increasing sales, checking product availability, including personalized recommendations, customer support, and order tracking and reducing operational costs, answer frequently asked questions (FAQs) and provide 24/7 customer support, which improves customer satisfaction and loyalty and enhance efficiency.

This chapter helps in giving a broader overview on Chatbots and XAI. It elaborates on the finer aspects of AI, ML, NLP, KG, Black Box, White Box, Attention Maps, Algorithms, Decision Tree, Random Forest, Linear and Logistic Regression. It introspects the nuances of Explainable AI and explores why XAI is becoming so immensely popular and revolutionizing E-commerce like never before. It also gives an overview of how XAI Chatbots are helping diverse industries to bring in a digitalized change in consumer service which is benefitting economies on a wider panorama.

2 Understanding Explainable AI (XAI)

As we all know that AI has already entered almost every facet of our life like, business, ecommerce, finance, healthcare, defence, education and many more, it is also increasingly important for all of us to also be assured that AI is making right decisions and its predictions are more accurate and correct as our trust and reliability on them is scaling great hights. [6] To make AI acquire that precision and transparency, Explainable Artificial Intelligence is a technique which enables us to understand AI and machine learning models, their predictions, accuracy, effectives, and decision making. It is extremely useful tool were high accuracy, stake decisions-medical diagnoses, financial matters, legal decision making, sports etc. As companies and businesses are now more accountable to maintain high quality standards, compliance, and data protection XAI serves as the most effective tool to maintain ambitious standards of trust and accuracy (Fig. 1).

Explainable artificial intelligence (XAI) is a unique technique in which advance set of procedures and processes are used that allows human users to comprehend and trust the results and output formed by machine learning algorithms efficiently and confidently. By Explainable AI and advance AI models one can easily analyse the expected impact and conclude the potential biases. It facilitates to define distinguish model accuracy, fairness, transparency, and productive outcomes in AI-powered decision making via a unique formulation. We can receive better productive results using Explainable AI as it helps organization in building trust and confidence. A company may adopt a responsible approach to AI development with the aid of AI explainability. Explainable AI facilitates the understanding of how efficiently Machine Learning (ML) and Artificial Intelligence (AI) work and what drives their decision-making [7]. With the advancing AI techniques, it is also becoming challenging for humans to interpret the complexity to comprehend and retrace how the algorithm is concluding

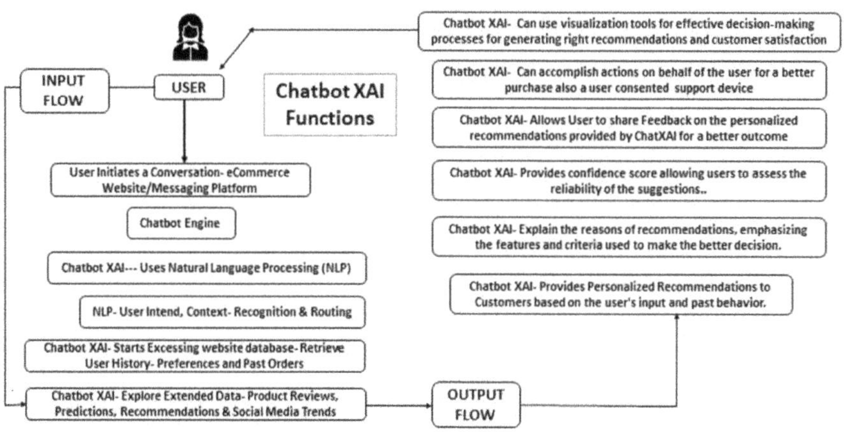

Fig. 1 Chatbots ecommerce functions

with precision and accuracy to a result. Many times, there is a sharp criticism whether the results outcome is dependable and rational.

The whole calculation process is turned into what is commonly referred to as a "black box" and how black box is calculating the interpretative outcomes that is impossible to interpret. Many ML systems, particularly deep neural networks, are regarded as black boxes because it is difficult to fully comprehend how they function once they have been trained. This makes it challenging to comprehend and justify a model's behaviour. By the help of explainability practisers can grasp better understanding how an AI assisted system works to generate a specific outcome. It allows the developers to confidently guarantee that the devices and systems are working as expected and must meet regulatory standards, or that those affected by a decision can challenge or change its consequences for a better outcome.

3 Importance of Explainable AI

For organizations adapting AI it is paramount to have deeper insights how AI works on decision-making algorithms with the use of model monitoring and accountability. XAI allows experts to comprehend and understand the effective functionality of machine learning (ML) algorithms, neutral networks, and deep learning. The decision-making of a Machine Learning models happens in such a manner that its input and operations are not visible to the user and third party. This configuration occurs in a most secret way where the user cannot get to see the inner functionality of the ML algorithms or clue how the results have been attained. This process is referred to black box, an AI formulation that are impossible to interpret and describe. Mostly secret electronic devices where human is lock up to understand and predict the operational outcome are called black box.

Explainable Artificial Intelligence refers to a technique of artificial intelligence system that helps in assisting and supporting customers to have a clear and logical explanations for a required question. It allows for its decision making and to adopt a right action. Chatbots AI and Natural Language Processing (NLP) system helps simulate human conversation and supports the business organisations in various purposes such as personal assistants, customer support language translation, real time engagement, and prevents grievance handling. Explainable AI in chatbots is especially important in today's scenario as Chatbots are used to collect customers queries in various segments. By the help of XAI Businesses and consumers will also get a deeper insight and understanding what are the reasons chatbot is supplying certain recommendations and supporting the consumer to take certain actions for a better outcome. As we also know chatbots are using advance machine learning algorithms with AI applications to develop user input to respond. XAI will help developers to navigate how chatbot is making decisions with the help of input.

In real sense XAI can be introduced into a chat bot for various purposes. By the help of a chatbot we can easily help user the reasons he should select a particular product or service. By the help of XAI a chat bot can also recommend or display various list of factors that can influence the decision-making process. Chatbot developers Can also take services like the decision trees on model interpretation tools for more exact transparent and explainable modelling. We can say, Explainable Artificial Intelligence (XAI) today plays a pivot role in advance machine learning, AI including chatbots and can help in enhancing real-time experience, engagement and most importantly trust and confidence between man and machine systems.

4 Chatbot

A chatbot is a well-constructed computer software program that uses artificial intelligence (AI) and Natural Language Processing (NLP). It is designed in a way that is capable of impersonating human conversations in both verbal and can effectively address various human personal and professional queries using natural tone, including text, or spoken language with the help of technologies like artificial intelligence (AI), Natural Language Processing (NLP), pattern recognition, etc.

4.1 XAI Chatbot and General Chatbot

XAI Chatbot and general chatbot both use Artificial Intelligence in their interactions with users [8]. There are some key differences that enhance the user experiences to the next level. [9]The difference between the two outlines what best augments your customer experience and work processes.

1. **Omni Communication Channels:** XAI can communicate explanations through voice as well, whereas general chatbot uses text commands to converse with users. XAI are more focussed on dialogues and other chatbots are more navigational based.
2. **Explainable in Nature:** As the name suggests XAI or Explainable AI are user friendly and give reasons for their end results. Since they are more transparent and have clearer communication traits, they are more popular than normal chatbots due to their confidence building features.
3. **Difference in Technology:** Improvised machine learning algorithms are used by XAI bots whereas traditional bots use rule based prior programmed systems which are not flexible. XAI analyses and learns from data to improve and makes them flexible to oversee complex tasks. XAI are non-linear with wider scope of interaction whereas general agent bots are set on canned linear interactions having limited scope to perform.
4. **NLP Levels:** XAI and general chatbots both use Natural Language Processing (NLP) for understanding and responding to queries. But advance form of NLP is used by XAI in comparison to a general chatbot. Entity recognition and sentiment analysis and other complex data analysis techniques are used by XAI for

more correct and personalized answers. XAI also uses Natural Language Understanding (NLU) simultaneously with NLP to better understand and interpret sentences in speech or text format. General chatbots rely more on prewritten scripts programmed to oversee basic information.

5. **Cost and Complexity:** Advanced and complex tools are used by XAI for better interface with clients. This makes them a little more expensive in implementation as compared to a normal chatbot. But once its implemented, it is well worth its name and advanced features. Its easy integration and deployment with existing text corpora and databases adds to its advantages, whereas a general chatbot has a complicated building process which is time consuming.
6. **Consumer Exposure:** XAI features are designed for an engaging and personalized user experience which reflects transparency, accuracy, and enhanced learning. Whereas there is a limited scope of response and communication retention in a generalized chatbot.
7. **Enhanced Responses:** A conversational chatbot like XAI can process dual requests and respond simultaneously at the same time. Whereas a general chatbot will revert to the first request, ignoring the second one during an interaction. Using advanced Data Science, ML, NLP helps XAI to judge voice and text inputs accurately to mimic and enhance conversational interaction flow flawlessly.
8. **Adaptable and Scalable:** XAI have a faster iteration with a regular updating and developing processes whereas in general chatbots reconfiguration is needed for any revision to set conversational flows and prior set rules. XAI is highly scalable due to its self-updating and evolving interface features, whereas other bots need regular updates, maintenance and revisions which are costly and time consuming.

4.2 XAI Concepts and Techniques

XAI terminology amalgamates understandability, comprehensibility, interpretability and explainability in one frame. This reflects to the attributes in a model made for human understanding about its functionality without detailing its algorithm and internal structure. The black box behaviour is characterised, and the models functional understanding is derived, which is further defined as understandability.

4.3 Explainable

XAI being explainable produces reasons with details to back its functioning and clear understanding of the users. Here methods are used to convert black box model interpretation, making it a post hoc (after this) interpretability. Model can be explainable, but its interpretability lies within the design of the model itself. XAI is outlined in two ways in academics and research and development. First that the models are transparent and interpretable by design. Second use of post hoc explainability or external techniques for explaining the models.

4.4 *Machine Learning Model Transparency Factors*

The degree of interpretability of a model is inbuilt in its characteristic and reflects its transparency [10]. Three major components defining the Model Transparency are:

1. **Simulatable:** When models can be thought and defined by humans with utmost ease, it is called simulatability. Such models should be self-confined for human minds to and think and engross upon it.
2. **Decomposable:** The model's ability to clarify every part of itself including parameters and inputs. Decomposability of a model is defined through its total interpretability by humans.
3. **Transparent Algorithms:** Algorithms are a set of rules and rigorous instructions used to solve problems or perform a computation. It also reflects the outline of how a model processes an input to give an output. Due to being algorithmic transparent, linear models error surface can be better understood and explained but it is not the same with deep learning architectures due to the loss landscape might be hazy.

4.5 *Linear, Logistic Regression, Decision Tree, and Random Forest*

Statistical Models are regularly used for predictive and classification study. [11] At times log odds are difficult to predict and understand, there some popular models within open end source tools, data science, Python, R which are incorporated to make the computations swift and easy.

1. **Linear Regression**

These models are used to classify the connection between one or more independent variables and a continuous dependent variable. In case there is one continuous dependent variable and one independent variable, it is termed as simple linear regression. When there are multiple independent variables, it is known as multiple linear regression. Through as set of data points, the best fitted line is plotted in every type of linear regression and the least squares method is used to calculate the same. Predicters are easily understandable. It is easier to recognize and can be done with a smaller sample size.

2. **Logistic Regression**

Logistic Regression is same as linear regression and is used to outline the connection between one or more independent variables and a continuous dependent variable with being used to make an estimation in relation to a categorical variable against a continuous one. So here, the classification model is used to predict a dependent variable as binary. Categorical variable can be yes or no, true, and false or 1 or 0 for that matter. As it reflects a profitability, the unit of measurability is also different from liner regression, whereas the s-curve is transformed into a straight line by logit function. The variables are easily readable and can be reasoned with easily. The models are transparent with their algorithms and standout the decomposability. Here a substantial sample size is required for value representation among the response class and to predict a noteworthy effect without sufficient statistical data.

4.6 *Decision Tree*

Decision Tree is an algorithm where supervised learning technique is used for classification and regression problems. It is called a classification algorithm as it is used in solving classification problems more. Here in a tree structured classification, where features of data set are represented by internal nodes and decision rules are represented by branches and further the outcome is represented by every single leaf node. The leaf node and decision node are the two main nodes of a decision tree. Decision nodes have multiple branches, enabling them to take decisions. On the other hand, Leaf Nodes do not contain branches, and are output of the decisions. The features of the given dataset function as a base for the decisions or test to be performed. The name decision tree is derived from its model similarity like a tree where a root node starts and expands to additional branches.

4.7 *Random Forest*

In Random Forest machine learning, algorithms are constructed and supervised from decision tree mechanism [12]. We see their application in e-commerce, banking, and other sectors to predefine outcome and behaviours. There can be many decision trees in a forest algorithm which is further trained and aggregated through bagging and bootstrapping. The precision of these machine learning algorithms is further improved by the cohesiveness of meta-algorithms known as bagging. The outcome of the predictions of the decision tree are established by the random forest algorithm. The precision of the outcome increases with the rise in the number of trees, as a mean average is derived from the number of involved trees. The limitations of a decision tree algorithm are reduced by random forest by reducing the overfitting of data sets, thus improving the precision ration. This does not need multiple configurations in packages (Fig. 2).

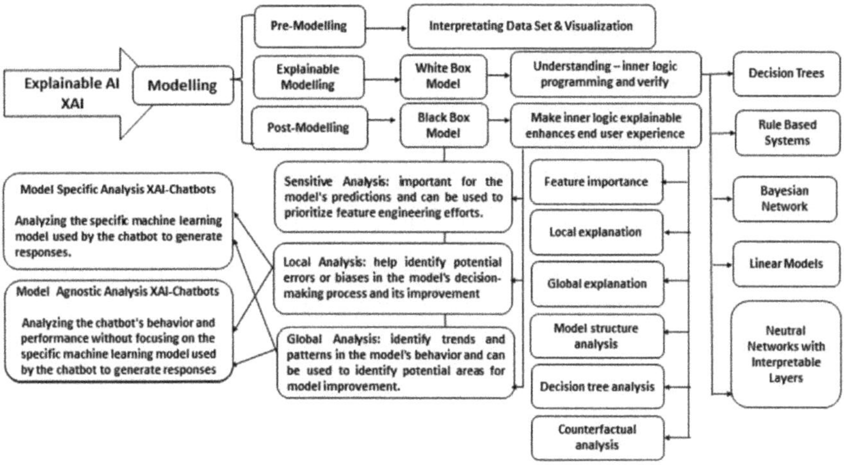

Fig. 2 Explainable XAI modelling

4.8 *Black and White Box Models*

XAI develops White Box models as well which are design explainable and require no extra features for interpreting explanations. On the other hand, Black Box models incorporate numerous techniques (Inner Logic/Output) to be explainable. White Box design models incorporates Handcrafted expert systems, Case based learning systems, Rule based learning systems, extracted systems, and Embedded symbols. Whereas Black box models combine with Model properties and Local and Global Logic to be more explainable. [13] The three stages of XAI are Pre-Modelling stage, Explainable Modelling stage and Post Modelling stage namely. Capability to explain to users why and how these predictions have been made is the core requirement of machine learning systems. So, when companies choose to integrate such ML models, preferring interpretability or accuracy is on top of their minds. [14]Black box models are more accurate with their gradient boosting models, neural networks, or intricate ensembles. Though their internal working is complex, there is also no estimation on each feature of the model prediction importance and how these diverse features interact and integrate. White Models are simpler and little weaker in comparison, with their decision trees and linear regression reflecting fewer predictive capability. They are often found lacking in outlining the complex datasets featuring interactions. However, they are easier to interpret and explain. While both are part of the data science continuum, black box reflects more machine learning whereas white box stresses on traditional statistics. Black box modelling has higher computational complexities, where focus is on speed and exploring a performant model while focusing on the quality of the prediction. White box modelling is simpler in its computation with focus on casual effects, processes, form, introspection while displaying a corrective model. Though outputs of black box AI are highly accurate but often the deep learning algorithms complexity leads to its lack of transparency which humans fail to understand. Their ground-breaking results are a mystery for the scientists and developers themselves. [15]White box models are gaining polarity with their transparency factors where scientist can predict their behaviour by having a look at the incorporated algorithm. They seem more practical for businesses as companies can pre understand their behaviour and act in accordance with using tangible to improve upon their existing workflows. White box insights are more linear and less disruptive and radical. Their predictions are not out of box or game changing but more dependable. Though they are not technically imposing but score on trust and reliability factors among users.

4.9 Numerous XAI Systems Can Be Enabled to Different Chatbots as Per Needs

1. **Rule based approaches:** Programming of Chatbots can be framed with a set sequence of rules formulated to respond to user input interface. By adding transparency and logical explanations, the chatbot outlines its outcome pattern and methodology.
2. **Decision Trees:** Designing of Chatbots can be done on decision tree format to map out the different conversation paths clearly. The decision tree enables the users to understand the route and explanation of a particular response.
3. **Attention Maps:** Attention Maps are used by Chatbots to highlight which parts of user's inputs where most applicable to the responses. Users then get to know, why the chatbot has provided this answer.

XAI tools can comprehensively enable chatbots with better understanding, logic, and transparency to improve overall user experience and improve the trust factors as an agent user interface between man and machine.

4.10 Chatbot Types

These are some of the different XAI enabled Chatbots which are classified and based on their diversified application functions. They use different techniques and programming as per explanation's provided by them on user interface demand.

1. **Dialog Explainable Chatbot:** They provide users with clarity to any explanations on different topics. The purpose is to function as a guidebook to diverse products and services. XAI enables further explanations to outputs for better understanding and follow up queries.
2. **Recommendation Chatbot:** Personalized suggestions are provided to users which are tailor made as per their past behaviour, understanding, and buying patterns.

3. **Transaction Chatbot:** These facilitate easy transactions such as making reservations and booking products. XAI helps by explaining why queries are being generated and how their transactions will be facilitated.
4. **Diagonstic Chatbot:** They identify and recommend solutions and remedies to technical, medical, and other issues and problems. XAI explains the diagnosis flowchart and why such remedies and solutions are being recommended.
5. **Feedback Chatbot**: Collection of feedback is generated by them for data analysis for products and services. XAI gives explanations to why certain set of questions are being asked and what will be the output purpose.

Many more diverse chatbots including explainable ones are need of the hour and can help in varied range of functionalities as per specific use and needs.

4.11 Beneficial Features of XAI Chatbot

XAI brings with it abundance of benefits which help industries to flourish. With proven outputs companies are becoming more effective by implementing XAI powered strategy and methodology [16].

1. **Interactive Customer Service:** XAI provides prominent level of customer service to clients through engaging interaction. It answers product and service specific answers, provides detailed information on features, and tries to resolve issues and problems. This saves energy, time, and resources by trying to be explainable in real time and building trust and transparency among clients.
2. **Individual Client Centric:** XAI offers one to one recommendation to customers based on their buying behaviour, pattern, and preferences. This adds to personalized user experience and client satisfaction.
3. **Reduces and Relieves Manpower:** XAI uses advanced automation techniques to resolve repetitive tasks such a scheduling appointment, booking and processing orders, collecting payments, managing deliveries, answering queries, and resolving them.
4. **Detects Irregularities Instantly:** XAI can quickly detect fraudulent transactions in real time to benefit clients and businesses. This helps authorities to take timely risk mitigation measures and avoid substantial losses.
5. **Accurate Analysis Projection:** XAI facilitates predictive analytics on consumer behaviour, buying pattern, trends, markets by compiling huge data banks. This data driven deep analysis helps organization in their futuristic growth plans and investments.
6. **Industry Driven:** XAI adapts to different industries specifically and is getting popular in Healthcare, Education, Finance, Marketing, Insurance, E-Commerce, and other sectors. In healthcare it is helping in booking appointments, providing patient care, monitoring patient health, and guiding in specific treatment, diet, and medication. In education field it provides remote education, personalized learning, answers queries and gives educational feedback to learners. It is helping

the financial and insurance sector in advising clients on savings, security, claims, conversions, frauds, planning, awareness, stability, markets, management, and the evolving economies. By maximizing ROI, providing strategies, building trust, and mitigating risks through accurate data analysis, it benefits the marketing sector immensely. It helps e-commerce in mapping production, storage, booking, delivery, payment, and other integrated client specific services. Other industries are also benefitting from their customer friendly services.

6. **Builds a Robust System:** XAI builds a transparent, logical databased ecosystem which enhances client's trust. This maps legal requirements with ethical integrations and justifies its robust and actionable deriving insights.

4.12 Challenges with XAI

XAI is an emerging and mass appealing notion as explanations to desired queries are a boon for customer service. At the same time satisfaction level, norms in societies and acceptance varies from individual to individual [17]. XAI similarly faces some challenges as far as policy making is concerned, which it looks to overcome with time.

1. **Lacking in Proficiency**: People may lack the expertise to access the explanations and agree to the outcome or fairness of the output.
2. **Data and Decision Dynamics**: Explanations can change as decisions and data can change overtime.
3. **Explanations can be Contested**: Algorithms explained by specialists differ or can be biased or at times they can make choices which are inherently debatable.
4. **Dependency on Context**: At a general level, algorithms cannot be explained or understood by everyone, as the outcomes might differ individually.
5. **Meddling of Algorithms**: In a complex scenario, there are many interlinked processes to collect and facilitate data from diversified sources using different algorithms.
6. **Addressal of Complex Problems**: Intricate problems are vague and hostile in structure. They can be solved in many ways as per the situation. Algorithm however provides one answer which can be arguable and change over a period.
7. **Decision Making is not based on Causality**: Causality should not be assumed as a source for decision making, though being used for as an explanation between input and output. Also, causality as explained might change from time to time.

4.13 Chatbots and Ecommerce

A well-constructed AI chatbot can effectively manage your customer 24×7, instant query by better engagement and providing personalized services by real time engagement, thereby helping your business save expenses, enhancing scalability, automate your lead generation and customer support, and personalize experiences for your customers. Chatbot uses natural processing language with artificial intelligence to interpret customers queries and simulates human conversation through a automate response. A special designed computer designed program is to reproduce human interaction with a computer machine to interact in a most conversational manner. Today chatbots are a big boom in the ecommerce sector as they are supportive in providing support and assistance to online consumers and shoppers. Today we can see chatbots giving assistance in websites, social media platforms and various texting applications devices.

4.14 Worth of Chatbots and Support Mechanism in Ecommerce

1. **Two Way Communication**: chatbots are highly effective in handling multiple inquiries related to a product or services. Online buyer can easily check his order status, track his order, return, or exchange policy. The time that a customer use to loss of customer care centre is now mitigating by real time response. It not only helps time but also reduces stress and burnout and builds trust and reliability.
2. **Personalized Marketing**: for a better Customer Engagement: chatbots can easily interpret the previous buying history, taste, and preferences. Since customers previous buying behaviour, the chatbots can effectively suggest better choice versions, comparative prizing, size availability etc. It can also help the investors to predict the future business layout based on footfall, choices, and trends. This way the investor can enhance his sales and profitable outcomes.
3. **Support Available 24 × 7:** Chatbots are computer operated tools thus can reach to diversified audience at the same time and various time zones. This method is more practical and helps the customers to even reach to the service provider at their time zones. Since the vast queries solutions are provided through a chatbot hence it helps in building customer loyalty and credibility.
4. **Tracking and Notifications**: Chatbots are very user friendly and provide lot of support and timely information. It helps customers to be abreast about the product order, its status, delivery time, return request and replacement statistics. In many ways chatbots help in a friendly way and reduces stress as its information is almost correct and it helps in navigating the trust of the buyer for a hassle-free purchase.
5. **Interactive Product Innovation and Engagement**: Chatbots can connect customers in collaborative and interactive two-way conversations to help them discover variety of products and services based on their preferences, budget, and needs. This can establish a further engaging shopping experience and enhances probabilities of customers finding products they are interested and engaged in.
6. **Response and Reviewing**: Time to time chatbots can solicit feedback and reviews from customer. This helps ecommerce to gather valuable feedback and insights on

their products, services, and after sales customer experience. Even the negative feedback helps the ecommerce to uplift and to improve products quality and standards. Doing this one can address customer concerns and enhance overall business outcomes.

4.15 Future of Chatbots in Ecommerce with XAI Facilitation

Chatbots are trending with technological interface systems powered by explainable AI and its growth in present and future looks immense. Ecommerce is a domain which is highly benefitting from XAI features as it provides enhanced consumer engagement for 24 h a day. Its real time responses with facts and data representation reduce response time, gives high retention rate, and automates routine tasks. With advancement of technology, chatbots are expected to become more advanced and enhanced in giving more personalized exposure to consumers. For example, with advanced language processing and machine learning interface, today chatbots acting as agents are interpreting clients queries better and reverting with very related and precise answers in real time.

XAI enabled chatbots are very crucial to the service industry as they are making technology facilitation easier and more accessible to the consumers. Having the unique ability to naturally interpret languages and revert with detailed responses makes them an absolute winner as far as AI interfaces with humans is concerned. Their logical and databased analysis further repose faith in their clients as to how they reached to a particular conclusion and further mobilises buyer patterns. Future of Ecommerce with XAI looks highly promising and will only increase by the day, with advancement of technology we can expect more consumer centric features in the coming days. Organizations and businesses are seeking efficient systems which are catering to customers in a personized, explainable, and accessible way and increasing sales and customer retention alongside.

Ecommerce facilitated by Chatbots are growing rapidly due to excellence in smooth consumer experience. These chatbots use NLP AND ML algorithms to interact with clients through conversations for online transactions while buying goods and services. Not only does this save time but enables high satisfaction to users. XAI incorporation techniques in chatbots also provides better transparency with complete accountability as it explains its logic behind the arrived conclusion. These AI explanations helps in building trust and provides customers a better insight into the machines decision making process.

4.16 XAI Revolutionizing E-Commerce

With Machine learning systems playing a greater, faster and vital role in deliverance of personalized customer care experiences in content and commerce, the client's buying exposure is changing 360 degrees. [18] These new age systems are offering marketers the option to exploit every penny spent on marketing programs through insightful action opportunities. XAI incorporates a Five model framework Answerable AI (AAI), Transparent AI (TAI), Impactful AI (IAI) Confident AI (CAI) with Responsible AI (RAI) and brings out revolutionizing exposure to the buyer and seller by bringing transparency, trust, ethics, and real time impact to the forefront in business decision making by being answerable and explainable. XAI serves as a true business value proposition as its explainability, business insights and data statistics helps improve business results and flawed systems on time which eventually saves money. For e-buyers XAI will provide reasons, explanations, expected outcome, quick decision making, model accuracy and learning engagement to enhance user experience in quick time frame in the most optimum way. This will increase product and service relevance, thus boosting consumers interest, confidence, engagement and buying ratio (Fig. 3).

By increasing trust factor and transparency, XAI helps in shaping the predicted conversion and reduction in abandoned shopping carts. It plays on the buying mindsets positively leading to more sales and higher revenue. The pandemic also changed the way people perceived e-commerce and technology, AI and MI made a game changing impact by having consumers amused through purchase of goods and services. During turbulence e-commerce helped many users also become entrepreneurs by initiating their home-based business. As per latest surveys, 95% consumers will use AI and MI in their digital interactions by 2025. Leaders in e-commerce are improving consumer experience through personalization, boosting sales of targeted clients, offering products on their prior buying, and browsing patterns, wishing people on occasions, reminding them of their wish list, personal reverts on enquiries and informing them on unique offers from time to time. [19]As per Gartner, the top priority of global CMO's is to invest in digital e-commerce, and 90% of them are aggressively investing in the same. AI is spreading its wing vigorously from Aviation to Cash Checkouts, and industry has seen around 270%

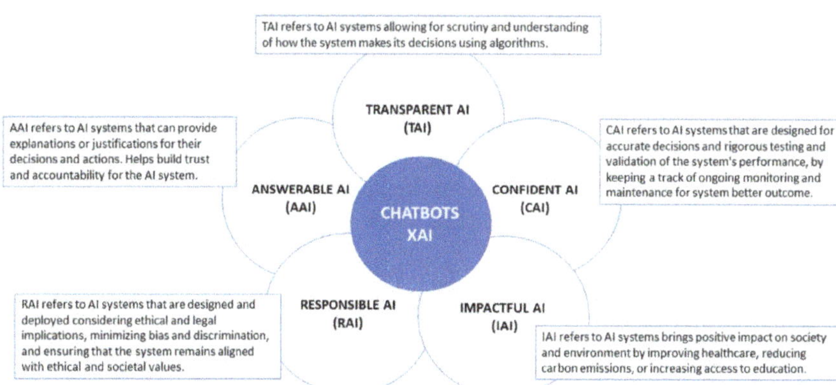

Fig. 3 The five model framework XAI chatbots

growth in last 4 years. Also, mobile users are using e-commerce for researching on products, services, comparing prices, offers and making online purchases. Interactive responses on product queries, prompt responses on digital touch points, personal suggestive preferences, lucrative offers, and secure pricing are foremost priority for most users. Present state of digital commerce has imbibed the concept of Understanding, Improving and Driving men and machines to integrate best practices in their businesses. XAI is helping diverse leadership teams to better outline digital mindsets by formulating data analysis, digital IQ, industry comparative charts to visualize performance graphs in fast evolving landscapes among diverse digital transformation options. XAI scores exceedingly well in overall performance of digital commerce maturity. Your revenue, customer acquisition, customer conversion, sales, profit, new channels and route to market and critical business capabilities all grow alarmingly. With notable applications like customized shopping, product proposal, chatbots, virtual assistants and voice search are adding to making a notable difference in consumer buying experience as far as personalized product recommendations, nuanced customer service, segregation of customers, smart logistics, forecasting of demand and supply is concerned [20]. Surveys have confirmed that in recent years, over 35-year-old in new consumers (60%) category have taken the lead in digital e-commerce buying against under 25-year counterparts.

4.17 *The Key Areas in E-Commerce Where XAI Can Be a Help Agent*

XAI is helping scientists and engineers to help explain business better for companies and clients [21]. Corporates and consumers always are looking to understand the big Why.

1. Facilitate Product (Clothes, Accessories, Glasses, Cosmetics, Etc.) Try on incorporating mirror technology (AR/AI) helping consumers with size, colours, fits, patterns, and designs digitally.
2. Static rules in CRM or Custom applications can be changed to dynamic customer journey.
3. Detecting rouge models, biases, and loopholes of skewed data promptly.
4. Consumer complaints and suggestions to be reviewed as per priority by highlighting their genuineness, stress, and pain points.
5. Identifying machine failures that are not subjected to core components of businesses.
6. Payments or Premiums not demarcated by clients Region, Gender, or Race.
7. Explaining every feature in detail for models to be used for credit scoring, thus helping in eliminating any prevailing inherent model bias.

4.18 *Major Hindrances in Incorporating XAI Enabled Chatbots in Ecommerce Industry*

It is crucial to understand the challenges or roadblocks XAI faces in its implementation [22]. So, its incorporation requires a well thought of plan to overcome the potential glitches [23]. As with any new change, that faces challenges and limitation issues, XAI must tide over these turbulences to achieve its objective.

1. **Understanding Consumer Mindset:** The major challenge with chatbots in e-commerce is getting an insight of the consumers mindset. Different clients have diverse needs and expressing ways through diversified dialects and languages. Understanding them and the language nuances is a challenge for these chatbots. Natural Language Processing (NLP) used in its much advance forms can be used to resolve the challenges of understanding the consumers mindset better through dialogue engagement processes or explainability [24]
2. **Building Trust through Transparency:** Trust and transparency are major concerns in e-commerce business. Clients can be extremely sensitive in sharing their personal and financial information with any agent or chatbot for that matter. So, its mandatory for chatbot designers to understand and program the bots that build transparency and trust factors through their prompt and explainable nature to dispense information desired by users. Consumers need to be told why a set of data is being collected from them, how it will be secured and what will be its end use.
3. **Data Security and Privacy**: Both private and public data is source of information for further analysis and decision making. Sensitive data handling is a matter of legality, guidelines, and protocol. The privacy of the sensitive data is immense and should be managed stringently and securely to avoid leakage by hackers or anti-social elements. The privacy of individuals and organizations should be managed at any cost with the most fool proof systems.
4. **Human Bias:** Apart from being transparent, any bias in data sources and algorithms used must be analysed carefully due to human error or system malfunctioning. Aggressive promotion of diversity and inclusion should be mandatory in AI system development.
5. **Personalised Attention**: Any buyer wants all the attention since he is the one spending on goods and services, so personalized attention is priority in e-commerce. The consumer is always apprehensive that the servicing team or chatbot might misuse his personal information gathered to better understand their choices and preferences. So, its mandatory that data protection and privacy is managed transparently and diligently by the organizations. Enhanced and assured personalised services can save time and retain consumers with real time positive buying exposure.
6. **Integrating Prevailing Systems:** With advancement of digitalization taking place quickly, ecommerce must integrate and upgrade existing systems with the trends. Inventory, warehousing, shipping, and deliveries systems must be in perfect sync. Seamless understanding of workflows and business processes

must be mapped with clarity and precision by both men and machines as they work in tandem.

7. **Data Quality Maintenance**: Data accuracy, completion, cleansing, enrichment, storage, governance, and monitoring are major concern factors for e-commerce. Poor data handling may lead to bias, incomplete lapses and security concerns which can hamper growth and reputation badly.
8. **Model Accuracy:** Use of correct model for different e-commerce business is vital. Thereafter feature selection, hyper parameter tuning, data volume training, model monitoring is equally important as XAI models can be complex.
9. **Cost Factor:** Efficiency in systems and services has a bearing cost in businesses including e-commerce. Though they bring numerous advantages of increased efficiency, decision making and overall real time experience. General cost includes data, development, maintenance, and implementation expenses.
10. **Black Box Model:** Machine learning systems based on black box models are complex and often difficult to predict. Often experts are baffled by the explanations based by the decisions of the algorithms. This can lead to hurdles in audit, verification, operation, legalization before implementation and guarantee thereafter. So, if there is a subpar analysis, at times it becomes difficult to analyse the poor output and how to correct them.

In a fast paced ever changing environment [25], XAI is incorporating new features in its robust technology for seamlessly interpreting and comprehending algorithms to overcome novel challenges.

4.19 XAI Chatbot Incorporation Techniques to Be Used in Ecommerce

1. **Personalized Interaction:** Chatbots as interface agents become an asset when they know their customers preferences and patterns through their past behaviour. This avoids repetition and quick facilitation of products and services.
2. **Algorithm Transparency:** Transparency is key to any business, so any algorithms used in digital transactions should be transparent. This reflects on the decision patterns and reliability of the chatbot.
3. **Explainable Nature:** Feature of being explainable as a chatbot is an asset, this helps in easy understanding of the outcomes and relates to consumers instantly.
4. **Judiciously Using Feedback:** Chatbots using feedback logically helps in the buyer's decision process and saves time immensely.

So XAI techniques enable chatbots to enhance the ecommerce business significantly making it more trustworthy, user friendly and enables personalized real time experience to consumers.

4.20 *XAI Demand, Supply and Surge Process*

Machine learning's tremendous success has enabled applications based on Artificial Intelligence to grow dramatically. There is a huge perception with positive promises of technological advance devices that will be autonomous in learning, deciding, perceiving, and taking decisions. Till now the satisfactory levels of these devices were hindered due to their inability to convince their decisions and answers to their users. The challenges of the industry are in lieu of expectations from users that more smart, independent and synergetic systems and devices originate which are more user friendly. Cater to this demand of a smart, user friendly chatbot who gives logical answers to his solutions, XAI was developed and presented to users. With the arrival of Explainable AI, the chatbot agents have added this advanced feature of explainable machine learning, which is easy to understand, trustworthy and reconciles data with accurate projections in advance. They function as firefighters to the upcoming igniting generation which is deeply engrossed in aligning with AI machine partners.

XAI backed by machine learning algorithms is a sequencing of processes programmed to be answerable with transparent result analysis output. It further describes AI as an explainable model with perceived impact and biases and further characterized on logic, patterns, accuracy, fairness, transparency, and results. Due to its trust building and confidence enabling features Explainable Artificial Intelligence becomes necessary for advanced organizations to produce as this reflects their accountable approach to the consumers and digital industry in general. With the advancement of AI, man and machine sync is growing real time. The black box manages the calculation system through data created by the algorithm used in a particular case. Developers and data scientists try to understand why the AI enabled system has given a specific output, this deep understanding helps them ensure that the machine is working as desired and meets the required regulations. If outputs are not as expected, then different algorithms are worked upon to change the previous undesired outcomes or probability to challenge the same.

Since XAI powered chatbots are becoming a rage, their adaptability is obvious and so they are being instantly incorporated by the leaders.

4.21 *Some of the Leading Brands Already Using XAI Enabled Chatbots Are:*

There are numerous innovative ways, how Brands and Businesses are incorporating chatbots to enhance user experience [26]. From buying clothes, booking flights, investments, education, schedules to recommendations, we are finding these explainable agents everywhere around us. They are finding new engaging tools to retain and delight consumers with their innovative features.

1. **Pizza Hut:** Pizza Hut Assistant is the brands chatbot that lets user select, order pizza and facilitates delivery through tracking. It uses its machine enabled algorithm to compile clients order and update regular delivery status.
2. **Starbucks:** My Starbucks Barista corresponds with customers to update orders with follow up questions through voice interaction and text messages both.
3. **Whole Foods:** This agent chatbot offers recipes to foodies using emojis. This culinary AI acts as a cookery expert for cooking enthusiasts and diet conscious patrons looking for healthy meal courses.
4. **H&M:** Kik by H&M advises customers on their preferences, colours, fits, and sizes for choosing the desired clothes. Here the agent chatbot uses Natural Language Processing tool to analyse past buying behaviour pattern and present choices and preferences to recommend best matching clothes as per their likes.
5. **Lyft:** Bot by Lyft will enable you to book a ride through messenger, slack, or voice. It helps in facilitating easy conveyance by sharing vehicle's current location, numberplate, and model.
6. **Wall Street Journal:** Its Facebook messenger chatbot alerts with stocks and big news in the market. Also enables customized reports, company analysis, financial metrics, and livestock quotes.
7. **Staples:** Uses its Intelligent Easy System chatbot to respond to order related customer questions. It also responds to stock in hand, tracking and handling returns.
8. **Spotify:** This Facebook messenger chatbot connects music lovers with searching, listening, and sharing music through playlist based on your moods.
9. **Fandango:** You can find local theatres, movie trailers, buy tickets and analyse trends on Fandango's messenger chatbot.
11. **Master Card:** XAI is used by Master Card to converse and help users in solving their queries relating to payments and other services. It understands the consumers queries and reverts with real time correct answers.
12. **Capital One:** Eno is the chatbot used by Capital One to function as their financial manager. It understands customers precise needs and personally recommends saving schemes and transactions relating to debt management.
13. **Sephora:** Chatbot Sephora Virtual Artist helps beauty product clients in using makeup products through augmented reality. This chatbot quickly analyses clients face and features to try, apply and recommend different product lines real time. They also allow clients to see how the products suits their face and persona before buying the actual products.
14. **Bizbike:** Bots from chatlayer hold interesting communications with e-bike clients and answers them reliably and efficiently.
15. **Belifus:** Insurance Claims are managed judiciously by the bank by these AI enabled chatbot. It helps Belifus from relieving stress of their staff.
16. **Yellow Class:** This interesting bot by Yellow Class handles volumes of educational queries, further forwarding complex queries to human agents.
17. **iFood:** Chatlayer chatbot by iFood enables food orders and delivery queries with ease and enhances customer support for drivers.

18. **European Commission:** Travellers to European Union are immensely benefitting in planning their journeys through the Bot, using its suggestions and guidelines.
19. **Nordstrom:** This Luxury fashion retailer uses chatbot 'Ask Nordstrom' to assist customers with products and style preferences.
20. **Bank of America:** Their Chatbot answers queries relating to banking, fund transfer and convenient bank and ATM locations.
21. **1–800-Flowers:** Customised Flowers, Gifts and deliveries are advised by the chatbot bringing numerous smiles to consumers.

Many other brands are similarly using XAI enabled user friendly chatbots to be answerable to users in real time and enhance their understanding transparently. This is also helping organizations to streamline operations and increase client satisfaction and retention significantly. With technology being developed constantly for better human machine interface, we see increased brands adopting XAI enabled chatbot agents as their key digitalized strategy to facilitate better and smoother client experience. Chatbots are helping them to stay a step ahead from their competitors by working smarter instead of working harder [27].

5 Conclusion

XAI is the new customer service norm with AI based solutions and are here to stay. A recent study by Price Water Cooper reflects that almost 52% companies have adopted conversational interfaces and digitalization of their organizations systems due to pandemic and incorporation of remote working lifestyle. Also, that further 86% of the respondents interviewed confirmed that AI is the mainstream technology running the businesses as on today. There is an imminent need to adopt and implement AI technology by those businesses who intend to stay ahead of the competitive edge. Increasingly customer-oriented organizations are using AI Chatbots, Answer Bots and XAI to improve Customer Satisfaction Index (CSI) and Customer Experience (CX) levels. Growth of Artificial Intelligence across diversified industries indicate the disruption levels of digitalization and technology prevailing in economies. XAI has breached everyday life of the average person making life convenient and well-versed. Earlier informed ideas were privilege of the elite class, technocrats, or data scientists only. With explainable AI even common consumers, executives, farmers, managers and government employees are all benefitting from handling transactions and queries with guidance on real time basis. [28] Though AI is decades old, researchers also paid fewer interest to XAI till year 2017–18 and the fact can be substantiated by limited research and publications on Explainable Artificial Intelligence till then. We can list out three major developments which highlighted and intrigued researchers towards the use of XAI from different domains using systems powered by AI/ML models towards its emergence as a technological servicing rage in the industry. First case was in early 2017 when Defence Advanced Research Projects Agency (DARPA) in USA started funding the Explainable Artificial Intelligence (XAI) Program. In a second breakthrough, 'The Development Plan for new generation of Artificial Intelligence' to motivate stronger and higher was backed by the Chinses Government in mid-2017. Lastly, the European Union in a major consumer friendly move enabled their citizens with 'Right to Explanation' by implementing the 'General Data Protection Regulation' (GDPR) if the citizens were anyhow impacted by any decision based on algorithm integration. Since then, no of research and development, publications, perspectives, and taxonomy about Explainable AI bloomed with new launches and adaption by diversified industries [29]. Even Google boldly announced their 'AI first' strategic policy in 2017.

Today data says, that most XAI technologies are being developed as Domain Agnostic [30] which lead the application domain race by approximately 48%, followed by most influential use in HealthCare by 22%, Misc. 7%, Industry 7%, Transportation 5%, Academics 3%, Judiciary 2%, Finance 2%, Entertainment 2% and Genetics 2%. If we review the application tasks percentage bifurcation their Decision Support leads the race with approximate 27%, Supervision 23%, Image Processing 21%, Recommendation System (10%), Predictive Maintenance (8%), Miscellaneous Tasks (5%), Business Management (3%) and Anomaly Detection (3%). As per experts globally AI technology will cross $4 Trillion by 2023 and more than $15 Trillion by 2030, expecting businesses to outline competitive models past promised results. Experts believe that XAI is the safe and structured path towards a responsible AI and an absolute need of the hour for businesses. Facts indicating its fastest growth among IT enabled technologies, truly substantiate its hyper automation success and rise. Scientists want them to be incorporated by industries as soon as possible due to their agility, speed and sophistication and being explainable at the same time. [31] Responsible AI (RAI) is the present and futuristic goal where transparency, customer orientation, ethics and regulations are foremost. [32]Their achievements at time have surpassed performances by humans in language translation, interpretation and image and speech recognition. With rapid changes in technology taking place every minute, we can foresee a novel generation of advanced systems and algorithms to avoid and replace prior issues with advanced XAI and RAI. With the capability to enlighten and educate users about their logic, rationale, strength and weakness characteristics, New Age Machine Learning systems are reflecting their futuristic behavioural pattern. XAI surely has a huge scope and very bright future in the machine human interface world.

References

1. Charmet, H. C. T. F. (2022). Explainable artificial intelligence for cybersecurity: A literature survey. Springer.
2. Sanke, A. (2022. April). What is explainable AI? 6 benefits of explainable AI. [Online]. Available: https://www.engati.com/blog/explainable-ai.
3. Saeed, C. O. W. (2023). Explainable AI (XAI): A systematic meta-survey of current challenges and future opportunities. Elsevier.

4. Menon, S. G. S. (2021, Dec 20). XAI is critical to ensure transparency. Analyticsindiamag.com.
5. Rajabi, K. E. E. (2022). Knowledge-graph-based explainable AI: A systematic review. Sage.
6. AKM, P. M. (2023). Explainable Artificial Intelligence (XAI) from a user perspective: A synthesis of prior literature and problematizing avenues for future research. Elsevier.
7. Laato, M. T. A. N. I. M. M. S. (2022). How to explain AI systems to end users: A systematic literature review and research agenda. Emerald.
8. Hyro. (2021, Jan 11). What's the Difference Between Chatbots and Conversational AI?. [Online]. Available: https://www.hyro.ai/post/whats-the-difference-between-chatbots-and-conversational-ai.
9. Bishop, C. (2023, April 20). Chatbots vs. conversational AI: What's the difference?. Zendesk Blog.
10. Venkataraman, V. A review of Explainable AI (XAI) concepts, techniques, and challenges. Deloitte.
11. IBM. What is logistic regression. IBM.
12. Mbaabu, O. (2020, Dec 11). Introduction to random forest in machine learning. [Online]. Available: https://www.section.io/engineering-education/introduction-to-random-forest-in-machine-learning/.
13. Guleria, M. S. P. (2022). Explainable AI and machine learning: Performance evaluation and explainability of classifiers on educational data mining inspired career counselling. Springer.
14. Hulstaert, L. (2019). Black-box vs. white-box models. *Towards Data Science*.
15. The Difference Between White Box and Black Box AI. [Online]. Available: https://bigcloud.global/the-difference-between-white-box-and-black-box-ai/.
16. Robison, R. Why Your Industry Needs Explainable AI (XAI) And How You Can Benefit From It. [Online]. Available: https://avianaglobal.com/why-your-industry-needs-explainable-ai-xai-and-how-you-can-benefit-from-it/.
17. de Bruijn, M. W. M. J. H. (2022). The perils and pitfalls of explainable AI: Strategies for explaining algorithmic decision-making. Elsevier.
18. Tulchinsky, A. Why Explainable AI (XAI) is the future of marketing and e-commerce. *The Future of Commerce*.
19. Gartner. Digital Commerce—Top Challenges, Trends and Proven Next Steps Strategy. [Online].
20. UNEXT. (2023 March 14). Artificial Intelligence and Machine Learning for E-Commerce. U Next.
21. Raju, V. Explainable artificial intelligence (XAI) is enabling data scientists to explain business better. [Online]. Available: https://enquero.com/explainable-artificial-intelligence-xai-is-enabling-data-scientists-to-explain-business-better/.
22. Echidna. Overcoming Challenges of Implementing AI in Ecommerce. [Online]. Available: https://echidna.co/blog/overcoming-challenges-of-implementing-ai-in-ecommerce/.
23. Espincorp. (2023, Aptil 12). Explainable AI: The Challenges and Limitation of XAI. [Online]. Available: https://www.e-spincorp.com/explainable-ai-the-challenges-and-limitation-of-xai/.
24. Meske, E. B. C. Explainable artificial intelligence: Objectives, stakeholders, and future research opportunities. Taylor and Francis.
25. Sonawane, N. (2022, June 5). Challenges of Explainable AI (XAI) and how to overcome them. Enterprise Talk.
26. Kim, L. (2018, Jan 5). Real Examples How Brands are Using Chatbot for Customer Service. [Online]. Available: https://medium.com/the-mission/10-real-examples-how-brands-are-using-chatbot-for-customer-service-4fbb5e4617f3.
27. Jayanti. (2023). Top companies using AI Chatbots to be a step ahead," https://www.analyticsinsight.net/.
28. Kale, T. N. A. (2023). Provenance documentation to enable explainable and trustworthy AI: A literature review. MIT Press Direct.
29. Fly, A. (2020). Is explainable AI (xAI) the next step, or just hype. *Towards Data Science*.
30. Mobyen. (2022). Theoretical background of explainable artificial intelligence. Scholarly Community Encyclopedia.

31. Chmiel, J. (2021). Explainable AI is what business needs in its path towards responsible AI. [Online]. Available: https://www.avenga.com/magazine/explainable-ai/.
32. E. A. S. R. J. N. I. A. P. M. A. Plamen P. Angelov. (2021). Explainable artificial intelligence: An analytical review. Wires.

Demystifying Applications of Explainable Artificial Intelligence (XAI) in e-Commerce

S. Faizal Mukthar Hussain, R. Karthikeyan, and M. A. Jabbar

Abstract The past ten years have witnessed significant advancements in artificial intelligence (AI), which has led to the widespread implementation of algorithms for the purpose of solving a wide range of issues. Nevertheless, in order to achieve this level of success, model complexity has been increased while at the same time black-box AI models, which lack clearness, have been utilized. With respect to the above mentioned demand, a concept known as XAI is presented with the goal of making AI more understandable and, as a result, accelerating the adoption of AI in important areas. Several relative ideas of XAI prime part in the published article, these difficulties and research ideas over Explainable AI are dispersed despite the fact that they have been identified. This chapter will provide an introduction to XAI, which will describe Why Explainable AI is needed, the various types of perspective groups, the three unique parts that make up XAI, and issues in XAI.

Keywords AI · XAI · E-Commerce · Financial applications

1 Introduction

It is not new problem faced in the name of the machine learning "black box" and prime reason is it takes over by many computational intelligence tasks: the requirement for openness, faith, and a proper knowledge of the future financial results. When making critical business decisions, it can be challenging and even dangerous to lack the knowledge necessary to connect the dots and predict the likely course of events. It's challenging to trust recommendations from a system which is yet to fully comprehend [1].

S. F. M. Hussain
Department of CSE, Mohamed Sathak Engineering College, Kilakarai, India

R. Karthikeyan (✉) · M. A. Jabbar
Department of CSE (AI&ML), Vardhaman College of Engineering, Hyderabad, India
e-mail: karthikhonda77@gmail.com

L. Gaur and A. Abraham (eds.), *Role of Explainable Artificial Intelligence in E-Commerce*, Studies in Computational Intelligence 1094,
https://doi.org/10.1007/978-3-031-55615-9_7

Several reasons is there for using XAI in financial applications. Being able to justify a recommendation or choice made by AI is an aspect that is becoming more and more crucial. Very next step of the future with respect to financial constraints is to integrate machines and humans. High level of transparency has been achieved by XAI which actually helping us the o have better focus over financial business and importantly helping us in taking decision at critical situation. It is also helping the admin to take decision over predicting future value. We have to clearly understand that XAI is not to eliminate human employees, but used to help them.

Finding the explanations for the advised action should be required whenever recommendations, advice, or even descriptive qualities are given. Without demonstrating the relationship to the data and the influencing elements utilized to arrive at it, it is not sufficient to merely predict the results or advise the best course of action. Key elements that influence predicted conversion will be provided by XAI. A business user will be able to make real-time adjustments to the important variables to further enhance profitability and company results. Due to this logic and transparency, there will be fewer discarded shopping heaves and a larger value in order value, which will increase returns and alteration rates. Any endeavor to improve the customer experience must incorporate the shift to Explainable AI, which is now under process. For taking business to the next level decision is made from the suggestion given by robots influenced by AI.

Additionally, XAI can contextualize the information for helping the any financial users to control the single action taken by them. Majority of the users utilize XAI to extract information and taking accurate decision over the circumstances led to time reduction [2]. If user doesn't find any help from the smart system then it will be big time consuming process and effort taking process. In the machine learning era, it is simple for common users to create their own AI solutions without understanding the details of how they operate. Because of this, it is challenging for business teams to use a reliable predictive or prescriptive model.

When developing predictive or prescriptive models, one of the major worries people have is that they don't understand how the computer makes decisions. It's crucial to remember that not only business users have trouble comprehending their models. This problem is also affecting ML engineers. A 2017 Mxnet poll found that 35% of AI professionals believed their colleagues did not understand how their models were created, and 47% felt uneasy sharing the results with them.

AI is always changing, and so are its applications. Currently, it assists e-commerce platforms in maintaining the happiness and amusement of their customers. In e-commerce, AI and ML are altering the game. The epidemic has also led to an increase in technology and online shopping. By 2025, 95% of all customer interactions will be handled by AI and ML, according to a survey. The leaders in the e-commerce industry are utilizing AI for e-commerce and ML to personalize the customer experience, target potential customers to increase sales, and provide products based on their browsing and purchasing habits. Personalized responses to customer enquiries, reminders about wish lists, and alerts about limited-time deals are some popular e-commerce strategies. From the increasing number of self-checkout cash registers to airport surveillance inspections, AI is present everywhere. A Gartner survey revealed

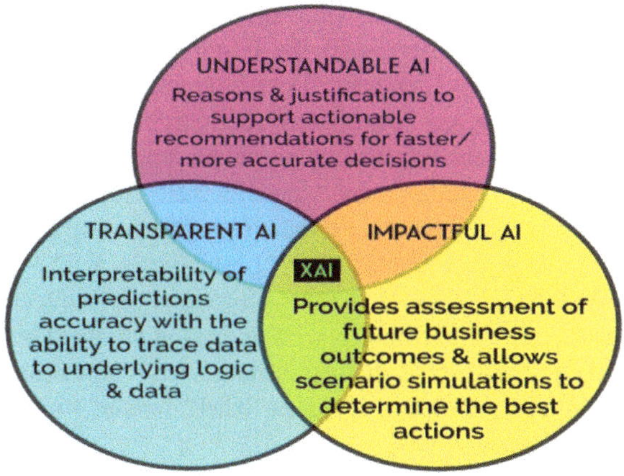

Fig. 1 Prime parts of XAI

that in the span of four years, the number of companies supporting AI is maximized. Nonetheless, there are a lot of myths about everything AI-related [3]. The history of AI should come first, and then we can discuss how it fits into the e-commerce industry.

2 Prime Parts of XAI

Figure 1 represents the three main parts of explainable AI. The parts are understandable AI, transparent AI and impactful AI. Understandable AI is used to take accurate decision, faster result, and every action decision support. Transparent AI is used to accurate prediction of future, tracing data and creating logic over data. Impactful AI is used to provide analysis over business environment and its data, its outcome and simulation based results for achieving better result in real time.

3 History of E-Commerce

Buying and selling products and services online is referred to as electronic commerce, or e-commerce. Many consumers regularly engage in e-commerce activities like shopping from internet stores or paying bills online. It encompasses a wide range of transactions and can take many various forms. In the UK in 1979, businessman Michael Aldrich developed the idea of internet shopping [3]. August 12, 1994, the "Internet Is Open, issue of the New York Times was fittingly named and narrates the operation of Compact Disc from one pal to other. As per the history mentioned by The Times, for achieving the anonymity fresh cyberspace owners have deal over the creating the readymade encryption software.

Figure 2 represent about the different stages of electronic commerce passes through various time periods. It starts from first online service provider to good ad words. The Electronic Data Exchange, developed in the 1960s, enabled electronic commerce. Electronic Data Interchange, often known as EDI, supplanted the traditional mailing and faxing of documents by facilitating data transfer between computers. Trade partners can interact electronically across industries in North America by exchanging orders, receipts, and other paperwork in a data format that complies with the ANSI ASC X12 standards. A VAN (Value-Added Network) checks an order after it has been sent before sending it to receiver system. The simultaneous movement of information with no person involved was made feasible with EDI. The websites built as "Hypertext project," with respect to a suggestion made in early 90 s by Tim and et al. The Dynatex SGML reader, which CERN licenced, served as the project's paradigm.

In 1991 Berners created the initial website server and tolls for using it. And he makes it free website for the public access. Because he wants to figure out the issues and problems arises while using the websites [4]. After NSF removed its limitations on NET business use in 1991, online revenues rose significantly. NSF star began

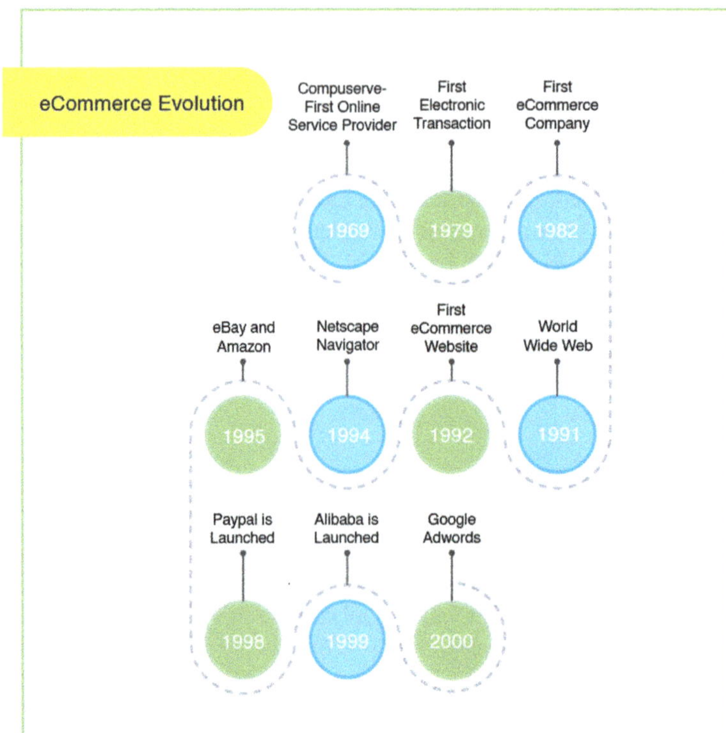

Fig. 2 Evolution of E-Commerce

collects to collect fee for domain registration by 1995. By 1993, there were two million domain names registered.

The introduction of two Coca-Cola vending machines in Finland in 1997 served as the first examples of mobile commerce. As more people began making purchases using their mobile devices and websites enhanced their user interfaces, mobile commerce accelerated over the next two decades. Mobile sales are predicted to make up 54% of all e-commerce purchases by 2021. Consumers and business buyers today use mobile devices to research products and find coupons, and social media usage is becoming more and more common. Corporate buyers need more consumer-focused features like personalisation and responsive design, as well as the ability to quickly access product information, secure pricing, and obtain online assistance.

4 Impacts of AI in E-Commerce

As more traditional businesses venture into the online retail space, artificial intelligence (AI) for e-commerce is becoming commonplace. In addition, it is boosting people's propensity to use and enjoy digital touchpoints. Integrating AI into your e-commerce website may increase sales while also enhancing productivity and efficiency.

4.1 Enhanced Target Marketing

In spite of the importance placed on personalization by the companies polled, only 15% assert that they have fully integrated it across all channels [5]. Thanks to developments in AI for online commerce, comprehensive personalization strategies are now feasible. By analyzing massive amounts of data from customers' purchase histories and other interactions, you can now better understand what they want and how to best deliver your message. Everything about your marketing strategy, from product development and branding to distribution channels, is informed by who you're trying to reach. AI can be a huge help in finding the right mood.

4.2 Stronger Bonds with Repeat Buyers

Sending clients individualized marketing and advertising communications helps increase retention. McKinsey found that adopting an omni channel approach to personalization increases revenue and customer retention by 10–15%. Superior data and insights about customers are a cornerstone of personalization, and their development is an asset that pays dividends at every stage of the value chain. The ROI on individualized marketing is projected to soon outpace that of traditional mass marketing [6].

4.3 Simple Mechanization

These days, you can shop online whenever, wherever across multiple platforms. This is typically a job that needs to be automated. By automating customer service, businesses can save money, resources, and effort. AI has the potential to greatly aid e-commerce business owners in automating the process of tracking and acting on marketing trends. E-commerce businesses can save time and money by using AI to automate tasks like encourage new goods, coordinating trade, identifying risky activity, encouraging faithful clients, and so on. Also, by letting chatbots handle the mundane questions, e-commerce store owners can focus on the more complex ones.

4.4 Efficient Sales Process

To their credit, most companies have moved on from antiquated sales practises such as scouring the reliable Yellow Pages and irritating potential customers by calling them cold. The lives of consumers are influenced in today's society by a variety of media, including commercials on television. In recent years Snapchat creates and shows the number of trendy opportunities over sales process. Combining AI and CRM is very needy because we want to have methods for solving the problem efficiently and for reaching the sales message to the clients. This is the best course of action if you want to achieve these goals. AI based electronic commerce has a variety of uses. Many artificial intelligence systems, such as Siri and Alexa, have the ability to learn natural language and take voice input. As a consequence of this, a customer relationship management system (CRM) may respond to customer inquiries and resolve customer issues. Some AI based CRM methods are capable of performing all of these responsibilities concurrently, in addition to a number of others [7, 8].

5 AI Uses in E-commerce

In the e-commerce industry, AI is a significant factor in the development of innovative solutions and improved customer experiences. The following is a list of the most notable applications of artificial intelligence in electronic commerce currently available:

1. Suggesting Product
2. Shopping Via Customized

3. AssistingVirtually
4. Chatbots
5. Searching Via Voice

In this context, most important applications of AI are as follows:

Providing Customized Recommendations for Products Artificial intelligence is used to make recommendations for individualized products based on the characteristics, such as behavior and appearance, of previous customers. Websites use a technique called machine learning to examine your shopping history and make product recommendations for you based on what you've bought in the past [9].

5.1 Personalized Attention to Each Client

You'll be able to provide superior levels of customer service by utilizing technologies like chatbots and virtual assistants. Even though they are not completely self-sufficient, these bots can streamline routine tasks, allowing live support organizations to focus their attention on more difficult inquiries. Additionally, virtual agents are available around the clock, which enables the immediate resolution of any issues that may arise.

5.2 Customer Segregation

With more data and processing power available from commercial and consumer sources, e-commerce firms may get complete realization of every customer and identify emerging trends.

Based on their findings, Accenture concludes that AI systems can rapidly evaluate a wide range of options for consumer participation and continuously enhance their performance as more data becomes available. As a result, businesspeople can adjust the AI's parameters to make it more precise.

5.3 Superior Arrangements

The foundation of smart and intelligent logistics is the use of sensors, RFID tags, and real-time information to better manage inventory and predict demand. Machine learning systems improve their predictions of logistics and supply chain activities as they gain in sophistication.

5.4 Trying to Guess Sales and Needs

This artificial intelligence method aids in evaluating and forecasting consumer interest in a product or service by analyzing historical data. Businesses can use demand forecasting to make more educated supply decisions based on an accurate projection of future sales and profits.

6 Main Perspectives for the Need for XAI

Now a day AI based on black box is used to get better result and decision over many areas including business. Decisions are taken without realizing anything over it. It is not mandatory for black box based AI to get proper explanation for each decision it receives from AI. Thus it reduces time complexity and deviation from its actual work. Figure 3 represents about the perspective of XAI. There are five perspectives mentioned in the picture. The perspectives are regulatory perspective, scientific perspective, industrial perspective, models development perspective and end user—social perspectives [10–13].

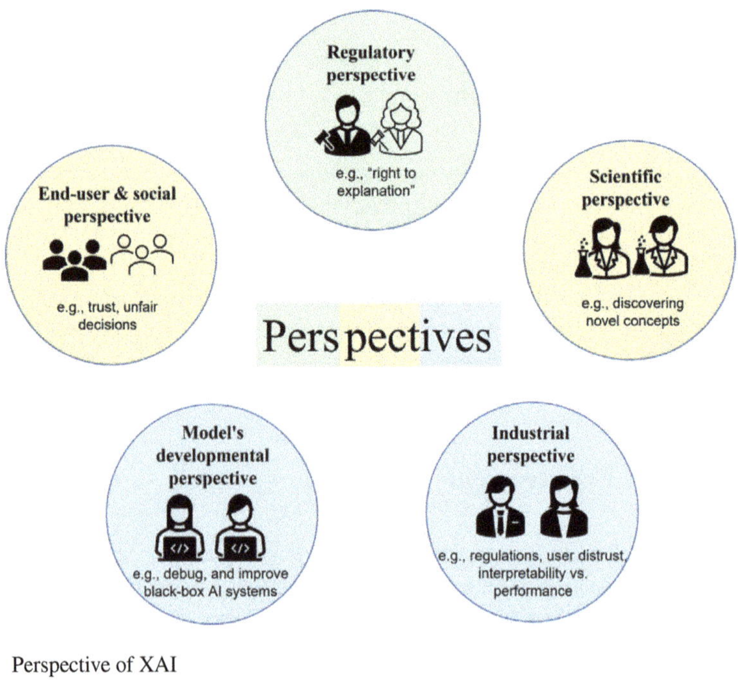

Fig. 3 Perspective of XAI

6.1 *Scientific Perspective*

The goal of developing a black-box AI model is to create a functional approximation that can solve the problem at hand. After developing AI methods, it becomes the source of information to a certain extent than the raw data. In light of this, XAI may be useful in elucidating the systematic awareness pull out AI methods, which in turn may show the way for the unveiling of previously unknown ideas in various scientific disciplines [12].

6.2 *Regulatory Perspective*

AI methods have huge impact over our daily activities, where it raises concerns from a regulatory standpoint because they may make decisions that are unacceptable, especially if they have legal consequences. As a result, it raises fresh concerns with the law. These rules establish the so-called "right to explanation," which grants users the opportunity to enquire into an algorithm's decision that has a significant impact on their lives. If AI methods deny a loan request, then the candidate has the right to know why and to make sure the system complies with all applicable laws and regulations [14]. However, implementing such rules is difficult, and the "right to explanation" is meaningless without enabling technology that can provide such explanations.

6.3 *Focusing on the Business World*

The industry faces obstacles in implementing highly difficult AI methods due to rules and user disbelieve in such classification. In the name of regulation, the industry may favour less accurate but more interpretable models. For example, XAI is useful because it can reduce the need to choose between a model's interpretability and its performance. However, it may raise expenses associated with production and rollout.

6.4 *Model's Developmental Perspective*

Lack of data needed to train, partial data, unwanted data, adversarial data, and model mismatch are all factors that could lead to inaccurate predictions when using black-box AI systems. Since black-box AI systems can have serious consequences for people's lives, it's important to comprehend what they've learned and why they make decisions. The goal is to use XAI to better comprehend and fix the AI methods so that it is more reliable, secure, and trustworthy to its users and exhibits less or no discriminatory, biased, or unfair tendencies. Further, XAI can aid selection when comparing models with comparable performance by disclosing the features used by the models to arrive at their conclusions. Since the algorithm may be fine-tuned for an insufficient objective, XAI can stand in as a surrogate for the real thing [15]. Forgetting about adherence rates while optimizing an AI system for cholesterol control is one example.

6.5 User and Societal Viewpoints

It has been demonstrated in the deep learning literature that fooling the model into assigning the wrong class label can be accomplished by manipulating an image in a way that is imperceptible to humans. In contrast, DL models can reliably recognize images that humans would have no chance of identifying at all. These results may cause some to question the reliability of black-box AI models. Another issue with opaque AI is that it has the potential to make biased judgments. This is a possibility if data exhibiting human biases and prejudices is used in the development of AI systems. In addition, interpretable models are often required because of the demand for the fairness of decisions made by black-box AI systems, which cannot be guaranteed by error measures. The aforementioned list is by no means exhaustive, and some of the aforementioned points of view may overlap. However, we think these viewpoints highlight the most important arguments for XAI's existence.

7 Problems with XAI

The new user coldstart problem is a challenge that has been encountered by many developers when working with collaborative filtering. According to Osman, the two biggest obstacles to the development of recommendation engines are cold-start and data sparsity. When a new user signs up for a system, they experience cold start, which means that the system is unable to make personalized recommendations to them right away because there are no data points to draw from.

Similar to cold-start, data sparseness refers to a user's lack of experience with an item, which leads to subpar recommendations. Still, Fernández-Tobaas and Braunhofer present three novel approaches in an attempt to solve this issue in their research paper. The recommendations made by "personality-based collaborative filtering" are better because they take into account the user's unique character. Personality-based active learning, which takes into account a user's unique traits in order to locate relevant data and resources. To compensate for a lack of user data in the target domain, "personality-based cross-domain recommendation" draws on user personality information from ancillary domains.

Because prior studies have shown that individuals with similar traits will also have similar preferences, their method centres on the user's personality. However, as is shown throughout the paper, one area of e-commerce that AI is having a major impact on is marketing. The use of AI in advertising can have a significant impact on consumer demand and purchasing behaviour. AI's perceived usefulness and perceived areas of use may mislead marketers and cause unrealistic expectations, despite the fact that AI provides decision makers with the data they need to overview areas of strength or deficiency [16].

Artificial intelligence (AI) systems are becoming more advanced, refined, and pervasive as this industry continues to attract investors; however, the processes for troubleshooting and supervising these systems are lagging behind their implementation. And because it relies so heavily on clicks and customer feedback, the data used by AI systems to analyse consumer behaviour is also unidirectional. Data of this nature, as stated by Khrais and Azizi, is both static and lacks depth, making it useless for gaining insight into consumers' habits as a whole. Although the difficulty in comprehending and defending AI system predictions has been around since the mid-1970s, recent advances in AI technology have focused on developing algorithms and models that produce more reliable forecasts. The reliability of these systems' forecasts, however, has not been a major concern until recently.

Due to their high complexity and autonomy, AI systems can develop a will of their own and make decisions with minimal or no human oversight. The widespread adoption of AI systems, and in particular machine learning, has led directly to this effect. Since existing systems don't provide in-depth explanations of how these models arrive at their results, predictions, or suggestions, the field of Explainable Artificial Intelligence (XAI) has seen a resurgence in recent years. Adabi and Berrada list the following four causes for the importance of XAI:

AI systems are not flawless and have historically provided users with discriminatory or biased results, so users will need to provide justifications. By incorporating XAI systems, users can provide justifications for the machine's unpredictable results or be assured that they will have a way to confirm that the algorithmic decision was fair and unbiased.

- Control Issues: The elimination of unfair judgements is aided by increased system control. Users who have a firm grasp on the system's inner workings are better equipped to spot vulnerabilities and fix them.
- Room for Development: Similar to the previous point, advanced users can identify vulnerabilities in the system. That's why it's great that XAI supports always getting better AI tools.

Desire for Knowledge: AI systems can help people learn new things if they are designed to be more accessible to humans.

These systems may one day be sophisticated enough to expand our understanding of fields like physics, biology, and more. In addition to these advantages, XAI will implement AI technologies in a more positive way because it will be easier to use and give its users more agency over the system. In addition, XAI addresses current ethical concerns, as existing systems learn from a wide variety of data, which can lead to inaccurate decisions. Therefore, more study of XAI can produce systems that are more transparent, reliable, and user-friendly. There has been a 250% increase in the amount of businesses using AI in recent years [17].

However, many companies have not realised AI's full potential because they do not know how to best leverage the supplementary resources that can improve the efficiency of these systems. Incompetence in integrating the system with the data and using the appropriate data to train AI is the primary problem, say Mikalef and Gupta. In addition, businesses need to realise that the entire process needs to be redesigned for automation to be beneficial; automating isolated processes will not produce the desired results.

In addition, "the ability of a firm to select, orchestrate, and leverage its AI-specific resources" is how Mikalef and Gupta characterise a company's AI Capability. Furthermore, they state that AI Capability can be further broken down into three distinct types of AI-centric resources: • Real a. Data: The accuracy of AI predictions and judgements is highly sensitive to the quality of the data used to train the system. Both internal (from departments like accounting and human resources) and external (from sources unrelated to the firm's operations) data are at a company's disposal. This means that businesses need to continuously refine the data used by these systems to ensure they achieve their desired outcomes. One of the biggest obstacles for businesses looking to adopt AI is the state of their technological infrastructure. In order to store massive amounts of data in multiple formats, businesses need scalable data storage.

The ability to quickly sift through massive amounts of data and process complex algorithms requires a company to have access to powerful computing resources. In order to get the necessary processing power, many businesses turn to GPU-intensive clusters, parallel computing, or the cloud. Companies must invest both time and money into AI projects in order to see the performance benefits they anticipate. • Human Businesses need a team of AI developers with the technical know-how to build and deploy AI algorithms, oversee the backend of their systems, and guarantee that their system is progressing in a way that serves the company's goals. e. Business Abilities: A lack of strong leadership is cited as the primary cause of a failed adoption of AI, even above a lack of technical knowledge on AI's applications.

Leaders need to see the potential in this technology's future and be able to steer their teams through transitions. • Intangible f. Coordination between departments It has been shown that inter-departmental coordination increases innovation within a company. Therefore, collaboration, shared objectives, and a shared vision are crucial components of a successful rollout of AI applications. g. Capacity for Organizational Change: Issues that may arise as a result of a shaky transition from the previous method to the new one are the focus of this consideration. Aversion to Risk: Many businesses may be hesitant to implement AI systems due to the novelty of the technology. On the other hand, those who aren't afraid to take chances will fall behind the competition and miss out on AI's benefits altogether. Many businesses that have adopted or are planning to adopt AI applications have found their weaknesses highlighted by these groups. But it also shows how focusing on complementary resources is likely to lead to an organization's superior performance after it has implemented AI technologies, which is a key takeaway from the article. However, CEOs should keep an eye on the tools used by their rivals, as disruptive technologies have the potential to render traditional approaches to doing business obsolete.

8 Deployment-Stage Difficulties and Future Research Priorities for XAI

In the sections that follow, we'll discuss some of the difficulties and potential areas of investigation associated with putting AI systems into action. Once ML solutions have been deployed, they will remain in use until they are no longer needed. The majority of the time, the provided explanations for AI systems are unchanging and only convey a single message. Understanding cannot be achieved through explanations alone.

Figure 4 represents about the list of challenges and research directions of XAI environment. Since the majority of existing XAI libraries lack user interactivity and personalization of the explanations, it is a promising research direction to advance the XAI field to allow users to explore the system via interactive explanations for a better understanding of the system. Several works have been proposed to improve human–machine teamwork beyond mere explanations [18].

Embodied agents with explicable agency. For autonomous agents to be considered "explainable," they must be able to provide justifications for their actions and the

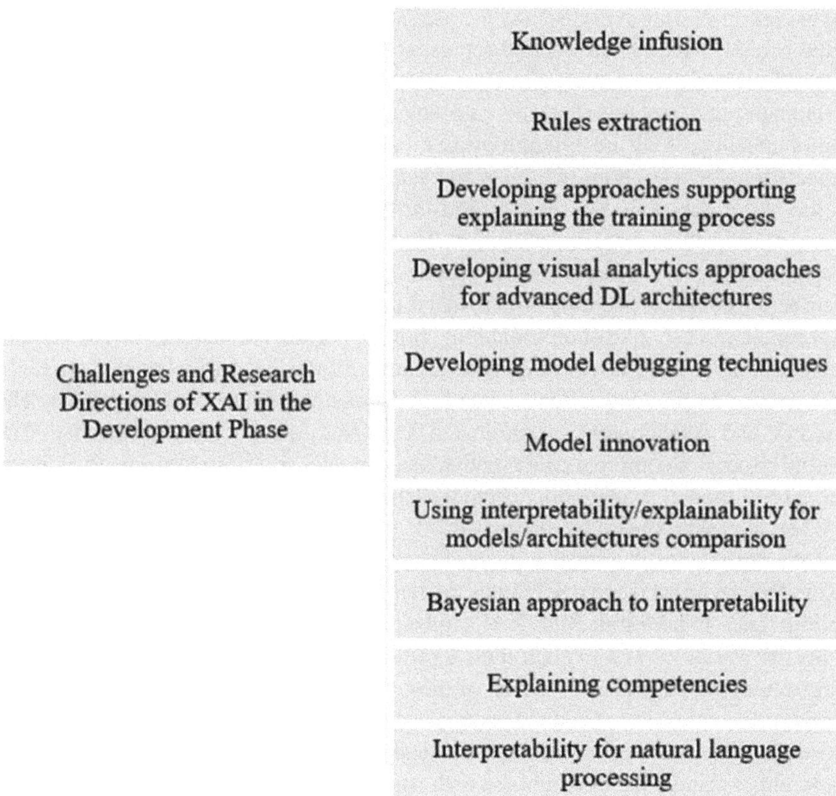

Fig. 4 Challenges and Research Directions of XAI

outcomes they intend to achieve. A proposed research roadmap for the explainable agency is based on the three stages of explanation.

The first step is to come up with reasons for why something happened or was accomplished. The following major lines of inquiry are being pursued during this stage of the research process: (1) The agent's or robot's built-in artificial intelligence system needs to be linked to its explanatory capabilities. (2) In order to create dynamic explanations, new mechanisms for locating pertinent explanation elements, determining their justifications, and combining them into a unified whole are required.

The second step is to exchange information that clarifies the situation. What information will be delivered to users and how it will be presented is the main concern here. To facilitate effective explainable agency communication, multimodal explanation presentations (e.g., visual, aural, and expressive) may prove useful in some circumstances. During the final stage, known as "explanation reception," the target is the human's comprehension of the provided explanations. There are things to think about to make sure the reception is accurate. Metrics should be developed to evaluate how well the explanations work and how users respond to them. Furthermore, the agent/robot should keep a model of user knowledge and keep updating it based on the evolution of user expertise and the user's perception of the agent/State robots of Mind (SoM), which is an internal representation of how the agent/robot treats the external world.

Humans and embodied agents can engage in verbal and nonverbal forms of communication. What are the appropriate models that can help to predict/track human expectations/beliefs about the goals and actions of an embodied agent? is just one of the many questions left unanswered in the literature on explainable embodied agents that needs further investigation. How can the Human-in-the-loop methodology be effectively incorporated into the design of embodied agents that can explain their actions? When deciding which social cues (such as words, gestures, or facial expressions) to use to explain something, how important are factors like context and embodiment? Is there a way to quantify trustworthiness? and why the explainability of AI has a complicated effect on the effectiveness of human-AI collaboration. The research and development of automated ML (AutoML) tools is on the rise. The AutoML tool is a complete data pipeline that can take you from raw data to an operational ML model. Explanation techniques that are model-agnostic can be used with any ML model produced by automated ML.

Similar to the explanation step, we can programmatically automate the following tasks: feature importance calculation, partial dependence plotting, surrogate model building, etc. In addition, Auto XAI can be programmed at a higher level to extract collective variables and explain their terms, such as by first extracting mathematical formulas used in the formation of the collective variables and then using these formulas to explain the generated predictions by ML. Existing AutoML tools include those that generate explanations mechanically. We anticipate that more Auto XAI tools, either standalone or combined with AutoML tools, or as services, will become available in the near future. Given that these would be provided as a service, it stands to reason that they would be designed to be extremely useful to a diverse group of consumers (e.g., non-technical experts) [19, 20].

Some notable factors we need to consider, according to the concept of research, include

- XAI and security
- XAI and reinforcement learning
- XAI and safety
- Machine-to-machine explanation
- Machine-to-machine explanation
- Explainable AI planning (XAIP)
- Explainable recommendation
- Explainable agency and explainable embodied agents
- XAI as a service
- Improving explanations with ontologies

We have covered topics like human–machine teaming, cross-domain artificial intelligence (XAI), security, issues and future research directions in XAI for reinforcement learning, XAI and safety, machine-to-machine explanation, and the two rights conflict in relation to the deployment phase (i.e., privacy and explanation). We also highlighted the importance of explaining AI in other contexts. Finally, we highlight the widespread use of Auto XAI as a service and the opportunity to enhance explanations through the use of ontologies.

9 Conclusion

AI has been producing good result regardless of the environment. It may be the better platform for electronic commerce. Our chapter has lot of suggestion and innovative ideas for electronic based activities. Also we present important applications of XAI apart from finance. Challenges and research areas related to XAI is addressed in all possible ways. Perspectives of XAI are mentioned in a proper way. While implementing XAI in electronic commerce what all are the impacts we need to face is also addressed.

References

1. Saeed, W., & Omlin, C. (2023). Explainable AI (XAI): A systematic meta-survey of current challenges and future opportunities. *Knowledge-Based Systems, 263*, 110273, ISSN 0950–7051. https://doi.org/10.1016/j.knosys.2023.110273.
2. Goodman, B., & Flaxman, S. (2017). European union regulations on algorithmic decision-making and a "'right to explanation.'" *AI Magazine, 38*(3), 50–57. https://doi.org/10.1609/aimag.v38i3.2741

3. Adadi, A., & Berrada, M. (2018). Peeking inside the black-box: A survey on explainable artificial intelligence (XAI). *IEEE Access, 6*, 52138–52160. https://doi.org/10.1109/ACCESS.2018.2870052
4. Gunning, D. (2016). Broad agency announcement explainable artificial intelligence (XAI). *Technical Report*.
5. Doshi-Velez, F., & Kim, B. (2017). Towards a rigorous science of interpretable machine learning. preprint arXiv:1702.08608.
6. Gade, K., Geyik, S. C., Kenthapadi, K., Mithal, V., & Taly, A. (2020).
7. Explainable AI in industry: practical challenges and lessons learned: implications tutorial. InProceedings of the 2020 Conference on Fairness, Accountability, and Transparency (FAT* '20).
8. Association for Computing Machinery. https://doi.org/10.1145/3351095.3375664.
9. Ahmad, M. A., Teredesai, A., & Eckert, C. (2018). Interpretable machine learning in healthcare. In *2018 IEEE-International Conference on Healthcare Informatics (ICHI)* (pp. 447–447). https://doi.org/10.1109/ICHI.2018.00095.
10. He, F., Wang, Y., Miao, X., & Sun, X. (2021). Interpretable visual reasoning: A survey. *Image and Vision Computing, 112*, 104194. ISSN-0262-8856. https://doi.org/10.1016/j.imavis.2021.10419.
11. Liu, M., Shi, J., Cao, K., Zhu, J., & Liu, S. (2018). Analyzing the training processes of deep generative models. *IEEE Transaction on Vision Computer Graphics, 24*(1), 77–87. https://doi.org/10.1109/TVCG.2017.2744938
12. van der Velden, B. H., Kuijf, H. J., Gilhuijs, K. G., & Viergever, M. A. (2022). Explainableartificial intelligence (XAI) in deep learning-based medical image analysis. *Medical Image Analysis, 79*, 102470. https://doi.org/10.1016/j.media.2022.102470
13. Orekondy, T., Schiele, B., & Fritz, M. (2019). Knockoff nets: Stealing functionality of black-box models. In Conference on Computer Vision and Pattern Recognition.
14. Huang, L., Joseph, A. D., Nelson, B., Rubinstein, B. I., & Tygar, J. D. (2011). Adversarial machine learning. In *Proceedings of the 4th ACM Workshop on Security and Artificial Intelligence, AISec '11, Association for Computing Machinery* (pp. 43–58). https://doi.org/10.1145/2046684.2046692.
15. Liu, S., Kailkhura, B., Loveland, D., & Han, Y. (2019). Generative counterfactual introspection for explainable deep learning. In *2019 IEEE Global Conference on Signal and Information Processing, Global SIP* (pp. 1–5). https://doi.org/10.1109/GlobalSIP45357.2019.8969491.
16. LeDell, E., & Poirier, S. (2020). H2O autoML: Scalable automatic machine learning. In *7th ICML Workshop on Automated Machine Learning, AutoML*. https://www.automl.org/wp-content/uploads/2020/07/AutoML_2020_paper_61.pdf.
17. Płońska, A., & Płoński, P. (2021). MLJAR: State-of-the-art automated machine learning framework for tabular data. Version 0.10.3, MLJAR, Łapy, Poland. https://github.com/mljar/mljar-supervised.
18. Panigutti, C., Perotti, A., & Pedreschi, D. (2020). Doctor XAI: An ontology-based approach to black-box sequential data classification explanations. In *Proceedings of the 2020 Conference on Fairness, Accountability, and Transparency*, in: FAT* '20, Association for Computing Machinery (pp. 629–639). https://doi.org/10.1145/3351095.3372855.
19. Confalonieri, R., Weyde, T., Besold, T. R., & del Prado Martín, F. M. (2021). Using ontologies to enhance human understandability of global posthoc explanations of black-box models. *Artificial Intelligence, 296*, 103471. https://doi.org/10.1016/j.artint.2021.103471.
20. Tudorache, T. (2020). Ontology engineering: Current state, challenges, and futuredirections. *Semant. Web, 11*(1), 125–138.

From Algorithms to Ethics: XAI's Impact on E-Commerce

Loveleen Gaur

Abstract "From Algorithms to Ethics: XAI's Impact on E-Commerce" explores the pivotal role that Explainable Artificial Intelligence plays in transforming the e-commerce landscape. It addresses the ethical challenges arising from algorithmic decision-making and underscores the significance of transparency, fairness, and trust in the world of online retail. As e-commerce continues to shape the future of commerce, XAI emerges as a fundamental enabler of ethical and sustainable digital engagement.

Keywords E-commerce · Explainable AI · XAI · Artificial intelligence

1 Introduction

In the ever-evolving landscape of e-commerce, the integration of artificial intelligence (AI) has become a defining feature, revolutionizing the way businesses engage with customers and optimize their operations. While algorithms have played a pivotal role in driving efficiency and personalization in online retail, they have also raised significant ethical concerns. This abstract delves into the transformative journey of e-commerce, from algorithms to ethics, through the lens of Explainable Artificial Intelligence (XAI) [1–8].

Artificial intelligence, driven by complex algorithms, has empowered e-commerce platforms to predict consumer preferences, recommend products, and streamline supply chain logistics. However, this era of algorithmic decision-making has also given rise to concerns regarding transparency and accountability. Customers often find themselves confronted with personalized product recommendations and pricing

L. Gaur (✉)
University of South Pacific, South Pacific, Fiji
e-mail: gaurloveleen@yahoo.com

Taylor's University, Subang Jaya, Malaysia

L. Gaur and A. Abraham (eds.), *Role of Explainable Artificial Intelligence in E-Commerce*, Studies in Computational Intelligence 1094,
https://doi.org/10.1007/978-3-031-55615-9_8

strategies, the inner workings of which remain hidden behind a digital veil. This opacity poses ethical challenges that demand attention and solutions.

Explainable Artificial Intelligence (XAI) emerges as a critical response to these ethical dilemmas in e-commerce. It represents the quest to make AI systems more transparent and interpretable, bridging the gap between algorithmic decision-making and human understanding. XAI seeks to provide users and stakeholders with clear insights into the factors influencing AI-generated recommendations and decisions [9–15].

At the heart of this transformation is the imperative of building trust. Trust is a cornerstone of successful e-commerce, and transparency is a prerequisite for trust. Shoppers need to understand why they are seeing certain products, why prices fluctuate, and how their data is used to shape their online experiences. XAI enhances this understanding by allowing users to peek behind the digital curtain, revealing the "why" and "how" of AI-driven recommendations. This fosters trust and loyalty in an era where consumer scepticism regarding data privacy and manipulation is on the rise.

Moreover, ethical considerations extend to the fair treatment of customers. AI systems can inadvertently perpetuate bias and discrimination when trained on historical data that reflects societal inequalities. XAI provides the tools to detect and rectify such biases. By making the decision-making process transparent, stakeholders can identify instances where AI may favour one group over another, enabling interventions to promote fairness and inclusivity in e-commerce.

The transition from algorithms to ethics also requires e-commerce platforms to grapple with the dualities of personalization and privacy. Consumers increasingly demand customized experiences, but they also desire control over their personal information. XAI can strike a balance by allowing users to set their preferences, understand how their data is utilized, and opt out of certain data-driven features while still enjoying personalized services. This negotiation between personalization and privacy is central to the ethical use of AI in e-commerce.

The impact of XAI on e-commerce reverberates beyond consumer interactions. It extends to supply chain optimization, fraud detection, and inventory management, areas where algorithmic decisions affect stakeholders such as suppliers and employees. XAI ensures that these decisions align with ethical standards, optimizing operations while respecting the rights and interests of all parties involved [16–20].

2 The E-commerce Revolution

2.1 The Digital Transformation

E-commerce, or electronic commerce, has revolutionized the way we shop, trade, and conduct business in the modern world. The rise of the internet, coupled with technological advancements, has ushered in an era where transactions, communication, and information exchange occur at the speed of light. This digital transformation has fundamentally altered the way businesses engage with customers, expand their reach, and optimize their operations. In this book, we explore how the integration of artificial intelligence (AI), specifically Explainable Artificial Intelligence (XAI), is a pivotal factor in this e-commerce revolution [21, 22].

2.2 The Algorithmic Foundation

The cornerstone of e-commerce's success has been the algorithms that underpin the industry. Algorithms are the complex sets of rules and calculations that power search engines, product recommendations, pricing strategies, and supply chain logistics. They drive efficiency, personalization, and responsiveness to customer needs. Algorithms have enabled e-commerce platforms to predict consumer preferences, streamline processes, and create seamless online experiences.

However, this era of algorithmic decision-making has not been without its challenges. It has raised significant ethical concerns, many of which revolve around the opacity of these algorithms. As customers engage with e-commerce platforms, they often find themselves presented with personalized product recommendations, dynamic pricing strategies, and curated shopping experiences. What lies beneath these decisions, though, remains a mystery to most consumers. This lack of transparency raises important questions [23].

2.3 The Ethical Dilemmas

The digital marketplace, characterized by its data-driven and algorithmic nature, has given rise to a series of ethical dilemmas that demand attention and solutions. As consumers, we want to know why we are seeing certain products, why prices fluctuate, and how our data is used to shape our online experiences. The imperative here is the need for trust. Trust is a cornerstone of successful e-commerce. In a world where scepticism about data privacy and manipulation is on the rise, transparency is a prerequisite for trust.

Moreover, ethical considerations extend beyond consumer interactions. They stretch into supply chain optimization, fraud detection, and inventory management, areas where algorithmic decisions impact stakeholders such as suppliers and employees. It becomes crucial to align these decisions with ethical standards and principles while optimizing operations [24, 25].

2.4 The Role of XAI

Explainable Artificial Intelligence (XAI) enters this landscape as a critical response to the ethical dilemmas inherent in e-commerce. It represents the quest to make AI systems more transparent and interpretable. XAI seeks to provide users and stakeholders with clear insights into the factors influencing AI-generated recommendations and decisions.

In this chapter, we delve into how XAI bridges the gap between algorithmic decision-making and human understanding, transforming the e-commerce landscape. It addresses the ethical challenges arising from algorithmic decision-making and underscores the significance of transparency, fairness, and trust in the world of online retail.

In the topics that follow, we will explore the impact of XAI on e-commerce, considering the transformative journey from algorithms to ethics. We will delve into transparency as a foundation of trust, the importance of fairness and bias mitigation, and the delicate balance between personalization and privacy. We will also examine practical considerations for implementing XAI in e-commerce platforms and provide case studies illustrating real-world applications. Finally, we will speculate on the future of ethical e-commerce with XAI, painting a picture of the developments and challenges that lie ahead [26–30].

3 The Algorithmic Era: Transforming E-commerce

Algorithms play a pivotal role in shaping the landscape of e-commerce. They underpin the core operations of online retail platforms and significantly influence the customer experience. Here are some key aspects of the role of algorithms in e-commerce:

- ***Product Recommendations:*** Algorithms analyse a user's past browsing and purchase history, as well as the behaviours of similar users, to generate personalized product recommendations. These recommendations are crucial for increasing sales and enhancing the user experience. By suggesting products that align with a customer's preferences, algorithms drive higher engagement and sales.
- ***Search Engines:*** Search algorithms power the search functionality on e-commerce websites. They help users find the products they're looking for quickly and efficiently by considering factors such as relevance, popularity, and user behaviour.
- ***Dynamic Pricing:*** E-commerce platforms often employ algorithms for dynamic pricing. Prices can fluctuate based on factors like demand, time of day, competitor pricing, and user behaviour. This allows businesses to optimize their revenue by adjusting prices in real-time.

- ***Inventory Management:*** Algorithms are used to optimize inventory management. They help businesses determine the right quantity of products to stock, considering historical sales data, seasonality, and other relevant factors. This minimizes overstocking and understocking issues.
- ***Supply Chain Optimization:*** Algorithms are instrumental in optimizing the supply chain. They help streamline logistics, minimize shipping costs, and reduce delivery times. These efficiencies benefit both businesses and customers.
- ***Personalization***: Algorithms are at the heart of personalized marketing and customer experiences. They enable businesses to target customers with specific offers, content, and promotions based on their preferences and behaviour. This level of personalization enhances customer engagement and loyalty.
- ***Fraud Detection:*** E-commerce platforms use algorithms to detect and prevent fraudulent activities, such as payment fraud and account takeovers. These algorithms analyse transaction patterns and user behaviour to identify suspicious activity.
- ***Customer Support:*** Chatbots and virtual assistants on e-commerce websites often rely on natural language processing algorithms to answer customer queries and aid. They can help improve the efficiency of customer support.
- ***Recommendation Engines:*** Beyond product recommendations, algorithms power recommendation engines for content like articles, videos, or music. By analysing user interactions and preferences, these engines suggest content that keeps users engaged and satisfied.
- ***A/B Testing***: E-commerce businesses use algorithms to conduct A/B tests to compare different versions of their websites, advertisements, or marketing campaigns. This data-driven approach helps optimize strategies for better results.
- ***Customer Segmentation:*** Algorithms segment customers into different groups based on various characteristics, including demographics, browsing behaviour, and purchase history. This segmentation allows businesses to target specific groups with tailored marketing strategies.

Algorithms are the backbone of e-commerce, enabling businesses to provide tailored, efficient, and personalized experiences to their customers. They drive efficiency, enhance user experiences, and optimize operations, all of which are essential in the highly competitive world of online retail. However, the increasing reliance on algorithms also raises ethical considerations, which has led to the development of Explainable Artificial Intelligence (XAI) to address transparency and fairness concerns in e-commerce [31–35].

4 Ethical Dilemmas in E-commerce

Exploring the ethical concerns arising from algorithmic decision-making is crucial in the digital age, where algorithms play an increasingly prominent role in shaping our lives. Here are some of the key ethical concerns that emerge from algorithmic decision-making:

- ***Transparency and Explainability:*** Many algorithms operate as "black boxes," making it challenging for users and stakeholders to understand the rationale behind their decisions. Lack of transparency can lead to mistrust and ethical issues, as individuals have a right to know how and why certain decisions were made that affect them.
- ***Bias and Discrimination:*** Algorithms can perpetuate bias and discrimination when trained on biased data. If historical data reflects societal prejudices, algorithms can unintentionally reinforce these biases. This can result in unfair treatment of certain individuals or groups, leading to social and ethical problems.
- ***Privacy Concerns:*** Algorithms often rely on vast amounts of user data, which raises privacy concerns. Collecting, storing, and using personal data without consent or in ways that individuals find intrusive can be ethically problematic.
- ***Autonomous Decision-Making:*** In autonomous systems, such as self-driving cars and AI-driven medical diagnoses, algorithmic decision-making can lead to life-or-death situations. Ensuring that these algorithms make ethical decisions that prioritize human well-being is a paramount concern.
- ***Unintended Consequences:*** Algorithms may produce unintended consequences or "side effects" that can harm individuals or society at large. For example, social media algorithms designed to maximize user engagement have been linked to issues like misinformation and filter bubbles.
- ***Accountability:*** Determining who is accountable for algorithmic decisions can be challenging. When things go wrong, it can be unclear whether responsibility lies with the developers, the data, the algorithms themselves, or other parties. This lack of accountability can hinder ethical oversight.
- ***Economic and Labor Issues:*** Automation and algorithmic decision-making in the workplace can lead to job displacement and economic inequality. Ensuring that these changes benefit society rather than exacerbating existing disparities is an ethical concern.
- ***Dehumanization:*** In some cases, excessive reliance on algorithms can dehumanize interactions. For example, chatbots and automated customer support systems may reduce the personal touch in customer service, which can lead to ethical concerns regarding the quality of service and the loss of jobs.
- ***Informed Consent:*** Ethical concerns arise when individuals are subjected to algorithmic decision-making without their informed consent. People should be aware of how their data is being used and the potential consequences of algorithmic decisions.

- ***Data Security:*** The security of data used by algorithms is of utmost importance. Data breaches and cyberattacks can have ethical implications, particularly when they compromise the privacy and security of individuals.
- ***Impact on Vulnerable Populations:*** Algorithms can disproportionately impact vulnerable populations, such as low-income communities, minority groups, or individuals with limited access to resources. Ethical concerns arise when these groups are unfairly disadvantaged.
- ***Environmental Impact:*** Algorithms that power resource-intensive processes, like cryptocurrency mining, can have a significant environmental impact. The ethical dimension of environmental sustainability is increasingly relevant in algorithmic decision-making.
- ***Data Ownership and Control:*** The question of who owns and controls the data used by algorithms is an ethical concern. Ensuring individuals have control over their data and receive fair compensation for its use is vital.

To address these ethical concerns, it is essential to develop and implement guidelines, regulations, and practices that promote transparency, fairness, accountability, and responsible AI development. The field of Explainable Artificial Intelligence (XAI) is one response to these concerns, aiming to make algorithmic decision-making more transparent and interpretable, thereby enhancing ethical oversight and mitigating potential harm. Ethical considerations should be central to the development and deployment of algorithms to ensure that technology serves the best interests of society [36–40].

5 Transparency as a Foundation of Trust

Transparency is a fundamental element in building trust with consumers, particularly in the context of e-commerce and businesses that rely on algorithmic decision-making. It involves open and honest communication, revealing the inner workings of operations, and providing clear, understandable information to customers. Here's why transparency is crucial for building trust with consumers:

- ***Predictability***: Transparency in business operations and algorithms makes outcomes more predictable for consumers. When users have insight into how decisions are made, such as product recommendations or pricing strategies, they can anticipate and understand the results. This predictability fosters a sense of control and reduces uncertainty, which is essential for building trust.
- ***Credibility and Reliability:*** Transparency is a sign of credibility and reliability. When companies openly share information about their processes, consumers are more likely to trust that the business is acting in good faith and providing accurate information. In turn, this trust can lead to increased customer loyalty.
- ***Informed Decision-Making:*** In transparent environments, consumers can make informed choices. They understand why certain products are recommended, how pricing is determined, and how their data is used. This empowers consumers to

make choices aligned with their preferences and values, enhancing their overall satisfaction.

- ***Mitigating Suspicion:*** Lack of transparency can fuel suspicion and skepticism among consumers. When they don't know why or how certain decisions are made, they may question the motives of businesses. By being transparent, companies can mitigate these suspicions and demonstrate their commitment to ethical practices.

5.1 Ethical Considerations

- Transparency is central to addressing ethical concerns. It allows consumers to evaluate whether a business's practices align with their values. For example, by revealing data usage policies, companies can demonstrate their commitment to respecting customer privacy and ethical data handling.
- ***Accountability:*** Transparent organizations are more likely to be held accountable for their actions. When consumers have insight into how decisions are made, they can hold the company responsible for any unfair or unethical practices. This accountability fosters a sense of justice and fairness, further strengthening trust.
- ***Brand Reputation:*** Transparency is closely tied to a company's reputation. Brands that prioritize openness and honesty are generally viewed more favourably by consumers. A positive reputation can lead to increased customer trust, loyalty, and word-of-mouth recommendations.
- ***Competitive Advantage:*** In a competitive marketplace, transparency can be a significant differentiator. Consumers are more likely to choose businesses that are open and clear about their practices over those that keep their operations shrouded in secrecy.
- ***Legal and Regulatory Compliance:*** In many industries, transparency is legally required. Failing to comply with transparency regulations can result in legal consequences and damage a company's reputation. Meeting these requirements is essential for maintaining consumer trust.
- ***Long-Term Customer Relationships:*** Building trust through transparency leads to long-term customer relationships. Customers are more likely to return to businesses they trust and recommend them to others. This can translate into sustained revenue and growth.

In the age of digital commerce, where personalization and data-driven decision-making are prevalent, transparency takes on added significance. Consumers want to know how their data is used, why they see certain content, and why prices vary. To meet these expectations and build trust, businesses must prioritize transparency, communicate clearly with their customers, and provide insights into the algorithmic processes that affect the customer experience [41].

Explainable Artificial Intelligence (XAI) is one tool that can help enhance transparency in algorithmic decision-making, making it easier for consumers to understand and trust the recommendations and choices presented to them. Ultimately, businesses that invest in transparency are more likely to cultivate lasting and trusting relationships with their customers.

6 Fairness and Bias in E-commerce

Fairness and bias in e-commerce are critical concerns that arise from the use of algorithms and AI in various aspects of online retail. Ensuring fairness and mitigating bias is essential to creating a level playing field for all customers and stakeholders. Here's a closer look at these issues:

Fairness in E-commerce: Fairness in e-commerce refers to the equitable treatment of all users and stakeholders, regardless of their demographic or personal characteristics. It involves creating an environment where everyone has an equal opportunity to access and benefit from online shopping and services. Fairness is crucial to building trust, maintaining customer satisfaction, and adhering to ethical principles. Several aspects of fairness in e-commerce include:

- *Pricing Fairness:* Pricing should be consistent and fair for all customers. Dynamic pricing strategies must consider factors like demand, costs, and competition but should not discriminate against specific groups or individuals.
- *Access to Products:* Everyone should have equal access to available products and services. This means ensuring that the allocation and distribution of products do not unfairly disadvantage certain groups or regions.
- *Customer Service Fairness:* Customer service should be provided consistently to all customers. It includes aiding, handling complaints, and resolving issues without bias or discrimination.
- *Promotion Fairness:* Promotional offers, discounts, and incentives should be applied fairly to avoid disadvantaging any customer group. Discriminatory promotion practices can lead to customer dissatisfaction and potential legal issues.

Ensuring that the user experience is designed to cater to diverse user needs, including those with disabilities, can enhance fairness and inclusivity [42].

Bias in E-commerce: Bias in e-commerce arises when algorithms and data-driven systems favour or discriminate against specific individuals or groups based on factors such as race, gender, age, or other characteristics. This can lead to unfair treatment and unequal opportunities for customers. The key concerns related to bias in e-commerce include:

- *Data Bias:* Algorithms often learn from historical data, which may contain biases present in society. If the data used for training algorithms is biased, the algorithms may inadvertently perpetuate these biases.
- *Recommendation Bias:* Recommender systems, which suggest products based on a user's history, may inadvertently steer individuals toward a limited range of choices, potentially reinforcing stereotypes, or biases.

- *Search Bias:* Search algorithms should provide users with diverse and relevant results, but they may inadvertently prioritize certain products or information over others, affecting users' choices and perceptions.
- *Pricing Bias:* Dynamic pricing algorithms could apply different prices to different groups based on historical spending habits, potentially leading to price discrimination.
- *Customer Service Bias:* Automated customer service systems may inadvertently treat customers differently based on language, accent, or cultural background, impacting the quality of service provided.
- *Content Bias:* Content recommendation algorithms, found in e-commerce sites and streaming platforms, can promote biased or polarizing content, influencing users' perspectives.

Addressing Bias and Promoting Fairness: To address bias and promote fairness in e-commerce, businesses can take several actions:

- *Diverse Data:* Ensure that the training data used for algorithms is diverse and representative of different groups, avoiding underrepresentation or overrepresentation.
- *Bias Audits:* Regularly audit algorithms to detect and mitigate bias. Bias detection tools and audits help uncover any discriminatory patterns.
- *Algorithmic Fairness:* Develop algorithms that are designed to be fair, considering fairness metrics and ensuring that decisions do not lead to discrimination.
- *User Feedback:* Encourage user feedback and complaints and respond to them promptly to rectify instances of unfair treatment.
- *Diversity and Inclusion:* Foster diversity and inclusion in the development and testing of algorithms and within the organization to reduce biases in decision-making.
- *Transparency*: Be transparent about the use of algorithms and data and explain how decisions are made to customers.
- *Legal Compliance:* Ensure compliance with anti-discrimination laws and regulations that may apply to e-commerce practices.

By addressing fairness and bias in e-commerce, businesses can create a more inclusive, ethical, and trustworthy environment for all customers, which, in turn, can lead to increased customer satisfaction, loyalty, and long-term success [43–45].

7 Balancing Personalization and Privacy

Balancing personalization and privacy are a delicate and essential consideration in the digital age, particularly in e-commerce, where companies aim to provide tailored experiences while respecting users' privacy. Striking the right balance between these two objectives is crucial for building trust with customers and complying with privacy regulations. Here's an exploration of this balance:

- ***Personalization***: Personalization in e-commerce involves tailoring products, content, and services to individual users based on their preferences, behavior, and historical interactions. It offers several benefits:
- ***Enhanced User Experience:*** Personalization can make the online shopping experience more convenient and enjoyable for customers by showing them products and content that align with their interests.
- ***Increased Engagement:*** Personalized recommendations and content are more likely to capture users' attention, keeping them engaged on the platform for longer periods.
- ***Higher Conversions:*** Personalized recommendations often lead to higher conversion rates, as users are more likely to purchase products, they find relevant.
- ***Brand Loyalty:*** Customers appreciate a personalized experience, and this can foster brand loyalty, leading to repeat business and positive word-of-mouth referrals.

Privacy: Privacy in e-commerce centres on safeguarding the personal information and data of users. Customers expect that their data will be handled responsibly and ethically. Privacy considerations include:

- ***Data Protection:*** Ensuring that user data is protected from breaches and unauthorized access is paramount. Security measures like encryption and access controls are essential.
- ***Informed Consent:*** Users should be fully informed about how their data is collected, used, and shared. Obtaining clear and explicit consent is a critical privacy practice.
- ***Data Minimization:*** Collecting only the data necessary for the intended purpose helps minimize risks and protect user privacy.
- ***Data Anonymization:*** Anonymizing or pseudonymizing data can protect individual identities while still allowing for useful data analysis and personalization.
- ***Compliance***: Adhering to data privacy regulations, such as GDPR in Europe or CCPA in California, is essential to avoid legal issues and maintain consumer trust.

Balancing the Two: Achieving the right balance between personalization and privacy involves several strategies:

- ***User Control:*** Allow users to control the level of personalization they receive. Give them options to adjust settings, provide feedback, and limit data sharing.
- ***Transparency***: Be transparent about data collection and use. Clearly communicate your privacy practices and inform users about how their data will be used to enhance personalization.
- ***Data Anonymization:*** When possible, use aggregated and anonymized data to create personalized recommendations instead of using individual data.
- ***Consent Management:*** Implement robust consent management tools that enable users to easily grant or withdraw consent for data collection and personalization.
- ***Data Retention Policies:*** Define clear data retention policies to limit the storage of user data, and regularly delete data that is no longer necessary.

- ***Regular Auditing:*** Regularly audit your data practices to ensure compliance with privacy regulations and to identify and address any potential privacy risks.
- ***Data Encryption:*** Use strong data encryption methods to secure user data, both in transit and at rest.
- ***Privacy by Design:*** Embed privacy considerations into the design and development of your e-commerce platform from the beginning.
- ***Continuous Monitoring:*** Continuously monitor for any changes in privacy regulations and adapt your practices accordingly.

Balancing personalization and privacy in e-commerce requires a thoughtful, proactive, and user-centric approach. By putting user privacy and data protection at the forefront of your business practices, you can build trust with your customers while still offering the benefits of personalization that enhance the overall e-commerce experience[46–48].

8 Implementing XAI in E-commerce

Integrating Explainable Artificial Intelligence (XAI) into e-commerce platforms is a complex task that requires careful planning and implementation. It is essential to ensure that AI systems provide transparent and understandable explanations for their decisions while maintaining or improving performance. Here are practical considerations and best practices for successfully integrating XAI into e-commerce platforms:

- ***Define Clear Objectives:*** Begin by defining the specific objectives and goals of integrating XAI into your e-commerce platform. Determine what aspects of the customer experience or business operations you aim to improve through XAI.
- ***Select Appropriate Use Cases:*** Identify use cases where XAI can have a meaningful impact. These might include personalized product recommendations, pricing strategies, search algorithms, or customer support.
- ***Data Quality and Preprocessing:*** Ensure that your data is of high quality, free from bias, and relevant to the use case. Preprocess and clean data as needed to enhance the effectiveness of XAI models.
- ***Model Selection:*** Choose XAI models that are suitable for the specific use case. Different XAI techniques, such as LIME (Local Interpretable Model-agnostic Explanations) or SHAP (SHapley Additive exPlanations), may be more appropriate for certain scenarios.
- ***Data Collection and Annotation:*** Collect user feedback and real-world data for training and testing XAI models. Annotate data with explanations to facilitate model training.
- ***User Interface and Experience:*** Design user-friendly interfaces that effectively communicate AI-driven decisions and explanations to users. Ensure explanations are presented in a clear, understandable, and non-technical manner.

- ***A/B Testing:*** Conduct A/B testing to assess the impact of XAI integration on key performance indicators (KPIs) such as conversion rates, customer satisfaction, and revenue. Compare XAI-enabled experiences with non-XAI experiences to evaluate the benefits.
- ***Privacy and Data Security***: Implement strong data protection measures to safeguard user data and maintain privacy. Ensure that XAI models are not exposing sensitive or personal information in their explanations.

8.1 Ethical Considerations

- Address ethical concerns, such as fairness and bias, when integrating XAI. Regularly audit and monitor XAI models to identify and mitigate any discriminatory or biased behaviour.
- ***Compliance with Regulations:*** Ensure that your XAI integration complies with relevant data privacy and AI regulations, such as GDPR, CCPA, and sector-specific guidelines.
- ***Scalability***: Design your XAI solution with scalability in mind. As your e-commerce platform grows, the ability to handle increasing data volumes and user interactions is crucial.
- ***User Education***: Educate users about the benefits of XAI and how to interpret the explanations provided. Clear user guides and support resources can help users understand and trust the AI system.
- ***Continuous Improvement:*** Maintain an iterative approach to XAI integration. Continuously collect feedback, monitor system performance, and update XAI models and explanations to adapt to changing user needs and preferences.
- ***Training and Support for Staff:*** Ensure that your team, including developers, customer support, and data scientists, is well-trained in XAI technologies and practices. This is vital for effective implementation and problem-solving.
- ***Feedback Mechanisms:*** Implement mechanisms for users to provide feedback on XAI-driven recommendations and explanations. This feedback can be valuable for improving the system over time.
- ***External Auditing:*** Consider external audits or third-party evaluations of your XAI system to ensure its effectiveness, transparency, and fairness.
- ***Documentation and Reporting:*** Maintain clear and detailed documentation of your XAI system's design, processes, and decisions, which can be useful for regulatory compliance and accountability.

By following these practical considerations and best practices, e-commerce platforms can successfully integrate XAI to enhance the user experience, improve decision-making, and ensure that AI-driven recommendations and decisions are transparent and trustworthy. This not only benefits users but also contributes to the ethical and responsible use of AI in e-commerce [49–51].

9 The Future of Ethical E-commerce with XAI

The future of ethical e-commerce with Explainable Artificial Intelligence (XAI) holds great promise and is poised to address many of the ethical concerns and challenges in the online retail industry. Here's a look at what the future may hold for ethical e-commerce with XAI:

- ***Enhanced Transparency:*** XAI will continue to evolve, providing even greater transparency into the decision-making processes of AI algorithms. Users will have clear, interpretable explanations for personalized product recommendations, pricing strategies, and other AI-driven features, fostering trust and confidence.
- ***Fairness and Bias Mitigation:*** XAI will play a central role in addressing issues of fairness and bias in e-commerce. Advanced algorithms will be designed to detect and mitigate bias, ensuring that recommendations, pricing, and customer interactions are free from discrimination.
- ***Regulatory Compliance:*** As data privacy and AI ethics regulations become more stringent, e-commerce platforms will increasingly rely on XAI to ensure compliance. Explainability features will assist in demonstrating adherence to legal and ethical standards.
- ***Customization and Control:*** XAI will enable e-commerce platforms to strike a balance between personalization and user privacy. Customers will have more control over the level of personalization they receive, adjusting settings to align with their comfort levels and preferences.
- ***Data Transparency and Consent***: E-commerce businesses will make significant efforts to ensure user data transparency and informed consent. Users will have clear visibility into how their data is used, and obtaining consent for data processing will be seamless and user-friendly.
- ***User-Centric Design:*** User interfaces will prioritize XAI explanations, making them an integral part of the user experience. Explanations will be designed to be easily understandable and accessible to a broad audience.

9.1 Ethical Retailer Alliances

- Retailers may form alliances and industry standards to promote ethical e-commerce practices. These initiatives could include shared XAI best practices and auditing standards to ensure transparency, fairness, and compliance across the industry.
- ***Data Trustworthiness:*** To prevent the use of poor-quality or biased data, e-commerce platforms will implement mechanisms to enhance data trustworthiness. Data sources will be carefully curated and regularly audited.
- ***Auditing and Accountability:*** Regular audits of XAI models will be conducted to ensure they are operating ethically. Businesses will be held accountable for any issues related to bias, privacy violations, or unfair practices.

- ***Responsible AI Development:*** E-commerce platforms will place a strong emphasis on the responsible development of AI and XAI. Ethical AI principles, such as fairness, transparency, and accountability, will be integrated into the AI development process.
- ***User Education and Empowerment:*** Users will become more informed about the benefits of XAI and their rights regarding data privacy and personalization. Educational resources and tools will empower users to make informed choices and exercise control over their online experiences.
- ***Sustainability and Ethical Supply Chains:*** XAI will be used to enhance the sustainability of e-commerce, addressing concerns related to environmental impact and ethical supply chain practices. Algorithms will optimize operations while adhering to ethical standards.
- ***Global Collaboration:*** Ethical e-commerce with XAI will transcend borders, with global collaboration and shared standards emerging as businesses seek to provide a consistent and responsible experience for users worldwide.

In the future, ethical e-commerce with XAI will not only address current challenges but also anticipate and adapt to emerging ethical concerns in the dynamic online retail landscape. It will align with the evolving legal and regulatory environment, prioritizing user trust, fairness, and transparency. This approach ensures that e-commerce remains a trusted and ethical domain for customers, fostering long-term relationships and sustainable growth [50–55].

10 Conclusion: Toward a More Ethical and Transparent E-commerce

The integration of Explainable Artificial Intelligence (XAI) into e-commerce holds the promise of creating a more ethical and trustworthy online retail environment. Transparency in AI decision-making is fundamental for building trust with consumers. XAI provides interpretable explanations for AI-driven recommendations and decisions, ensuring users understand and trust the process. Also, striking a balance between personalization and user privacy is crucial. E-commerce platforms must empower users to control their personalization preferences while safeguarding their data and privacy. Ethical e-commerce with XAI actively addresses bias and discrimination in algorithmic decision-making. Advanced algorithms detect and mitigate bias, ensuring that recommendations, pricing, and customer interactions are fair and non-discriminatory. As data privacy and AI ethics regulations become more stringent, ethical e-commerce with XAI ensures compliance with legal and ethical standards, protecting both consumers and businesses. User interfaces are designed with XAI explanations in mind, making them easily understandable and accessible. Explanations become an integral part of the user experience, fostering user trust. Collaboration among retailers and industry standards promote ethical e-commerce practices. These alliances aim to ensure transparency, fairness, and compliance across the industry. Businesses invest in mechanisms to enhance data trustworthiness, curating data sources and conducting regular audits. Accountability is prioritized, and businesses are held responsible for ethical concerns. Ethical AI principles, such as fairness, transparency, and accountability, are integrated into the AI development process. Ethical considerations are a fundamental part of AI technology. XAI is used to enhance the sustainability of e-commerce by optimizing operations and adhering to ethical standards. The path forward for ethical e-commerce with XAI involves continuous improvement, adaptation to evolving ethical concerns, and a commitment to ensuring a responsible and trustworthy online retail environment. It is a journey that aligns with the changing legal and regulatory landscape, prioritizing user trust, fairness, and transparency. This approach ensures that e-commerce remains a trusted and ethical domain for customers, fostering long-term relationships and sustainable growth in the digital marketplace.

References

1. Gielens, K., & Steenkamp, J. B. E. (2019). Branding in the era of digital (dis) intermediation. *International Journal of Research in Marketing, 36*(3), 367–384.
2. Tan, F. T. C., Pan, S. L., & Zuo, M. (2019). Realising platform operational agility through information technology–enabled capabilities: A resource-interdependence perspective. *Information Systems Journal, 29*(3), 582–608.
3. Khrais, L. T., & Azizi, T. A. (2020). Analyzing consumer attitude toward mobile payment technology and its role in booming the e-commerce business. *Talent Development and Excellence, 12*.
4. Areiqat, A. Y., Hamdan, A., Alheet, A. F., & Alareeni, B. (2021). Impact of artificial intelligence on E-commerce development. In *The Importance of New Technologies and Entrepreneurship in Business Development: In The Context of Economic Diversity in Developing Countries: The Impact of New Technologies and Entrepreneurship on Business Development* (pp. 571–578). Springer International Publishing.
5. Gururaj, P. (2021). Artificial intelligence-application in the field of e-commerce. *Int. J. Res.-Granthaalayah, 9*, 170–177.
6. Firschein, O., Fischler, M. A., Coles, L. S., & Tenenbaum, J. M. (1973, August). Forecasting and assessing the impact of artificial intelligence on society. *IJCAI, 5*(1), 105–120.
7. Soni, N., Sharma, E. K., Singh, N., & Kapoor, A. (2020). Artificial intelligence in business: From research and innovation to market deployment. *Procedia Computer Science, 167*, 2200–2210.
8. Shyna, K., & Vishal, M. (2017). A study on artificial intelligence E-commerce. *International Journal of Advances in Engineering & Scientific Research, 4*(4), 62–68.
9. Chaudhary, M., Gaur, L., & Chakrabarti, A. (2022, April). Comparative analysis of entropy weight method and C5 classifier for predicting employee churn. In *2022 3rd International Conference on Intelligent Engineering and Management (ICIEM)* (pp. 232–236). IEEE.
10. Rana, J., Gaur, L., Singh, G., Awan, U., & Rasheed, M. I. (2021). Reinforcing customer journey through artificial intelligence: A review and research agenda. *International Journal of Emerging Markets, 17*(7), 1738–1758.
11. Chaudhary, M., Gaur, L., Chakrabarti, A., & Jhanjhi, N. Z. (2023). Unravelling the Barriers of Human Resource Analytics: Multi-Criteria Decision-Making Approach. *Journal of Survey in Fisheries Sciences*, 306–321.
12. Khrais, L. T. (2020). Role of artificial intelligence in shaping consumer demand in E-commerce. *Future Internet, 12*(12), 226.

13. Afaq, A., Gaur, L., Singh, G., & Dhir, A. (2021). COVID-19: Transforming air passengers' behaviour and reshaping their expectations towards the airline industry. *Tourism Recreation Research*, 1–9.
14. Van Lent, M., Fisher, W., & Mancuso, M. (2004, July). An explainable artificial intelligence system for small-unit tactical behavior. In *Proceedings of the national conference on artificial intelligence* (pp. 900–907). AAAI Press; MIT Press, 1999.
15. Gunning, D., Stefik, M., Choi, J., Miller, T., Stumpf, S., & Yang, G. Z. (2019). XAI—Explainable artificial intelligence. *Science robotics*, *4*(37), eaay7120.
16. Barocas, S., Friedler, S., Hardt, M., Kroll, J., Venka-Tasubramanian, S., & Wallach, H. (2018). The FAT-ML workshop series on fairness, accountability, and transparency in machine learning. *cit. on*, 7.
17. Afaq, A., Gaur, L., & Singh, G. (2023). Social CRM: Linking the dots of customer service and customer loyalty during COVID-19 in the hotel industry. *International Journal of Contemporary Hospitality Management, 35*(3), 992–1009.
18. Wilburn, K. M., & Wilburn, H. R. (2018). The impact of technology on business and society. *Global Journal of Business Research, 12*(1), 23–39.
19. Statista. Future Areas in Which AI Will Help E-Commerce Businesses According to Decision Makers in North America and Europe in 2021. 2021. Available online: https://www.statista.com/statistics/1175063/ai-help-business-ecommerce-companiesworldwide/ (Accessed on 10 November 2021).
20. Chaudhary, M., Gangele, A., Naved, M., Gaur, L., & Singh, G. (2022, November). The function of driver categorisation in the ride-hailing industry: A study on on-demand transport. In *2022 3rd International Conference on Computation, Automation and Knowledge Management (ICCAKM)* (pp. 1–6). IEEE.
21. Statista. AI Use Cases in Consumer Goods and Retail Industry Worldwide as of 2020. 2020. Available online: https://www.statista.com/statistics/1197958/ai-use-cases-consumer-goods-retail-global/ (accessed on 5 December 2021).
22. Anshu, K., Gaur, L., & Singh, G. (2022). Impact of customer experience on attitude and repurchase intention in online grocery retailing: A moderation mechanism of value Co-creation. *Journal of Retailing and Consumer Services, 64*, 102798.
23. Afaq, A., & Gaur, L. (2021, November). The rise of robots to help combat covid-19. In *2021 International Conference on Technological Advancements and Innovations (ICTAI)* (pp. 69–74). IEEE.
24. Khalid, O., Khan, S. U., & Zomaya, A. Y. (Eds.). (2019). *Big Data Recommender Systems: Algorithms, architectures, big data, security and trust* (Vol. 1). Institution of Engineering and Technology.
25. Sorbán, K. (2021). Ethical and legal implications of using AI-powered recommendation systems in streaming services. *Információs Társadalom: Társadalomtudományi Folyóirat, 21*(2), 63–82.
26. Sharma, S., Singh, G., Gaur, L., & Afaq, A. (2022). Exploring customer adoption of autonomous shopping systems. *Telematics and Informatics, 73*, 101861.
27. Lundberg, H., Mowla, N. I., Abedin, S. F., Thar, K., Mahmood, A., Gidlund, M., & Raza, S. (2022). Experimental analysis of trustworthy in-vehicle intrusion detection system using explainable artificial intelligence (XAI). *IEEE Access, 10*, 102831–102841.
28. Cirqueira, D., Nedbal, D., Helfert, M., & Bezbradica, M. (2020). Scenario-Based Requirements Elicitation for User-Centric Explainable AI: A Case in Fraud Detection. In *Machine Learning and Knowledge Extraction: 4th IFIP TC 5, TC 12, WG 8.4, WG 8.9, WG 12.9 International Cross-Domain Conference, CD-MAKE 2020, Dublin, Ireland, August 25–28, 2020, Proceedings 4* (pp. 321–341). Springer International Publishing.
29. Yang, G., Ye, Q., & Xia, J. (2022). Unbox the black-box for the medical explainable AI via multimodal and multi-centre data fusion: A mini-review, two showcases and beyond. *Information Fusion, 77*, 29–52.
30. Cirqueira, D., Helfert, M., & Bezbradica, M. (2021, July). Towards design principles for user-centric explainable AI in fraud detection. In *Artificial Intelligence in HCI: Second International*

Conference, AI-HCI 2021, Held as Part of the 23rd HCI International Conference, HCII 2021, Virtual Event, July 24–29, 2021, Proceedings (pp. 21–40). Springer International Publishing.
31. Liao, Q. V., & Varshney, K. R. (2021). Human-centered explainable ai (xai): From algorithms to user experiences. arXiv preprint arXiv:2110.10790.
32. Rodrigues, V. F., Policarpo, L. M., da Silveira, D. E., da Rosa Righi, R., da Costa, C. A., Barbosa, J. L. V., Antunes, R. S., Scorsatto, R., & Arcot, T. (2022). Fraud detection and prevention in e-commerce: A systematic literature review. *Electronic Commerce Research and Applications*, 101207.
33. Adadi, A., & Berrada, M. (2018). Peeking inside the black-box: A survey on explainable artificial intelligence (XAI). *IEEE access, 6*, 52138–52160.
34. Brendel, A. B., Mirbabaie, M., Lembcke, T. B., & Hofeditz, L. (2021). Ethical management of artificial intelligence. *Sustainability, 13*(4), 1974.
35. Ekhart, N. (2022). Taking down malicious webshops: Designing Explainable AI against growing e-commerce fraud.
36. Yılmaz Benk, G., Badur, B., & Mardikyan, S. (2022). A new 360° framework to predict customer lifetime value for multi-category e-commerce companies using a multi-output deep neural network and explainable artificial intelligence. *Information, 13*(8), 373.
37. Chaudhary, M., Gaur, L., & Chakrabarti, A. (2022, November). Detecting the employee satisfaction in retail: A Latent Dirichlet allocation and machine learning approach. In *2022 3rd International Conference on Computation, Automation and Knowledge Management (ICCAKM)* (pp. 1–6). IEEE.
38. Matuszelański, K., & Kopczewska, K. (2022). Customer Churn in retail e-commerce business: Spatial and machine learning approach. *Journal of Theoretical and Applied Electronic Commerce Research, 17*(1), 165–198.
39. Waltl, B., & Vogl, R. (2018). Increasing transparency in algorithmic-decision-making with explainable AI. *Datenschutz und Datensicherheit-DuD, 42*(10), 613–617.
40. Hulsen, T. (2023). Explainable Artificial Intelligence (XAI) in Healthcare.
41. Ferreira, J. J., & Monteiro, M. S. (2020). What are people doing about XAI user experience? A survey on AI explainability research and practice. In *Design, User Experience, and Usability. Design for Contemporary Interactive Environments: 9th International Conference, DUXU 2020, Held as Part of the 22nd HCI International Conference, HCII 2020, Copenhagen, Denmark, July 19–24, 2020, Proceedings, Part II 22* (pp. 56–73). Springer International Publishing.
42. Suresh, A., & Rani, N. J. (2020). Consumer perception towards artificial intelligence in E-commerce with reference to Chennai city, India. *Journal of Information Technology and Economic Development, 11*(1), 1–14.
43. Islam, A. S., Ahmed, S., & Khan, R. H. (2022, March). A review on e-commerce system in Bangladesh: An empirical study. In *Proceedings of the 2nd International Conference on Computing Advancements* (pp. 269–276).
44. Bradley, C., Wu, D., Tang, H., Singh, I., Wydant, K., Capps, B., Wong, K., Agostinelli, F., Irvin, M., & Srivastava, B. (2022, November). Explainable artificial intelligence (XAI) user interface design for solving a Rubik's cube. In *HCI International 2022–Late Breaking Posters: 24th International Conference on Human-Computer Interaction, HCII 2022, Virtual Event, June 26–July 1, 2022, Proceedings, Part II* (pp. 605–612). Springer Nature Switzerland.
45. Antoniadi, A. M., Du, Y., Guendouz, Y., Wei, L., Mazo, C., Becker, B. A., & Mooney, C. (2021). Current challenges and future opportunities for XAI in machine learning-based clinical decision support systems: A systematic review. *Applied Sciences, 11*(11), 5088.
46. Ravi, M., Negi, A., & Chitnis, S. (2022, April). A comparative review of expert systems, recommender systems, and explainable AI. In *2022 IEEE 7th International conference for Convergence in Technology (I2CT)* (pp. 1–8). IEEE.
47. Ahmed, I., Jeon, G., & Piccialli, F. (2022). From artificial intelligence to explainable artificial intelligence in industry 4.0: A survey on what, how, and where. *IEEE Transactions on Industrial Informatics, 18*(8), 5031–5042.

48. Gaur, L., Afaq, A., Singh, G., & Dwivedi, Y. K. (2021). Role of artificial intelligence and robotics to foster the touchless travel during a pandemic: A review and research agenda. *International Journal of Contemporary Hospitality Management, 33*(11), 4079–4098.
49. Wolofsky, S. (2020). What's your privacy worth on the global tech market? weighing the cost of protecting consumer data against the risk that new legislation may stifle competition and innovation during this global, technological revolution. *Fordham Int'l LJ, 44*, 1149.
50. Gaur, L., & Sahoo, B. M. (2022). Intelligent transportation technology enablers. In: Explainable Artificial Intelligence for Intelligent Transportation Systems. Springer. https://doi.org/10.1007/978-3-031-09644-0_2.
51. Gaur, L., Ratta, M., & Gaur, A. (2022). Future of DeepFakes and ectypes In: Deepfakes. CRC Press, 9781003231493.
52. Gaur, L., & Sahoo, B. M. (2022). Explainable AI in ITS: Ethical concerns. In: Explainable Artificial Intelligence for Intelligent Transportation Systems. Springer. https://doi.org/10.1007/978-3-031-09644-0_5.
53. Gaur, L., & Sahoo, B. M. (2022). Intelligent transportation system: Modern business models. In: Explainable Artificial Intelligence for Intelligent Transportation Systems. Springer. https://doi.org/10.1007/978-3-031-09644-0_4.
54. Gaur, L., Jhanjhi, N. Z., Bakshi, S., & Gupta, P. (2022). Analyzing consequences of artificial intelligence on jobs using topic modeling and keyword extraction. In 2022 2nd International Conference on Innovative Practices in Technology and Management (ICIPTM) (pp. 435–440). https://doi.org/10.1109/ICIPTM54933.2022.9754064.
55. Gaur, L., Bhandari, M., Razdan, T., Mallik, S., & Zhao, Z. (2022). Explanation-driven deep learning model for prediction of brain tumour status using MRI image data. *Frontiers in Genetics, 13*, 822666.

GPSR Compliance
The European Union's (EU) General Product Safety Regulation (GPSR) is a set of rules that requires consumer products to be safe and our obligations to ensure this.

If you have any concerns about our products, you can contact us on

ProductSafety@springernature.com

In case Publisher is established outside the EU, the EU authorized representative is:

Springer Nature Customer Service Center GmbH
Europaplatz 3
69115 Heidelberg, Germany

www.ingramcontent.com/pod-product-compliance
Ingram Content Group UK Ltd.
Pitfield, Milton Keynes, MK11 3LW, UK
UKHW022209051026
15624UKWH00001B/82
9783031556142